Palgrave Studies in the Future of Humanity and its Successors

Series Editors

Calvin Mercer
East Carolina University
Greenville, NC, USA

Steve Fuller
Department of Sociology
University of Warwick
Coventry, UK

Humanity is at a crossroads in its history, precariously poised between mastery and extinction. The fast-developing array of human enhancement therapies and technologies (e.g., genetic engineering, information technology, regenerative medicine, robotics, and nanotechnology) are increasingly impacting our lives and our future. The most ardent advocates believe that some of these developments could permit humans to take control of their own evolution and alter human nature and the human condition in fundamental ways, perhaps to an extent that we arrive at the "posthuman", the "successor" of humanity. This series brings together research from a variety of fields to consider the economic, ethical, legal, political, psychological, religious, social, and other implications of cutting-edge science and technology. The series as a whole does not advocate any particular position on these matters. Rather, it provides a forum for experts to wrestle with the far-reaching implications of the enhancement technologies of our day. The time is ripe for forwarding this conversation among academics, public policy experts, and the general public. For more information on Palgrave Studies in the Future of Humanity and its Successors, please contact Phil Getz, Editor, Religion & Philosophy: phil.getz@palgrave-usa.com.

More information about this series at
http://www.palgrave.com/gp/series/14587

Debra J. Bassett

# The Creation and Inheritance of Digital Afterlives

## You Only Live Twice

palgrave
macmillan

Debra J. Bassett
The Centre for Death & Society (CDAS)
The University of Bath
Bath, UK

Palgrave Studies in the Future of Humanity and its Successors
ISBN 978-3-030-91683-1 ISBN 978-3-030-91684-8 (eBook)
https://doi.org/10.1007/978-3-030-91684-8



Cover credit: Yuichiro Chino / Getty Images

This Palgrave Macmillan imprint is published by the registered company Springer Nature Switzerland AG.
The registered company address is: Gewerbestrasse 11, 6330 Cham, Switzerland

*This book is dedicated to Trevor William Greenwood (1935–1986),*
*my greatest inspiration.*

# Preface

Nine years ago, my friend's daughter Annie, aged 18, died suddenly. At the time of her death, Annie had an active Facebook page where she regularly posted photographs of friends and family. Even today, Annie remains 'socially alive' on her live Facebook page. Following Annie's death, I became interested in why my friend—and many others—still spoke to her on Facebook in the present tense as though she were reading the posts. At the time, Facebook allowed the memorialisation of pages following the death of the user. However, some inheritors of these pages—who have access to the accounts—chose not to do this, enabling their deceased loved ones to remain socially active in a digital afterlife.

Some tech companies are moving into the Digital Afterlife Industry (DAI) using intelligent algorithms that enable us to create digital doppelgängers of us when we are alive, which they claim could continue to interact with the world following our biological death. These technologies provide new ways for us to communicate with the dead and foster ongoing relationships in our digitally mediated societies. Following initial research on digital inheritance I began to wonder if this type of posthumous digital endurance would affect how people grieve; moreover, I wanted to understand if these digital artefacts of the dead were experienced differently from physical keepsakes by those who inherit them. My Dad had died over 30 years earlier, and I began to think about how it would feel to possess digital memories and messages from him and whether this would have been a comfort or disruption to my grief. I thought this ethnographic study had begun; however, when I told Katie, my daughter, about the idea, she told me it was not serious research—rather, it was an episode of Black Mirror that she had recently watched. At this point, I had not heard of the Netflix series written by Charlie Brooker, but with the wind removed from my sails, I sat down to watch Be Right Back. Rather than disheartening me, this episode—which explored the reanimation of the dead—succeeded in further igniting my curiosity about the ramifications of this type of human-computer interaction.

Until recently, the dead used to reside with … well … the dead, in cemeteries or dusty photograph albums where more of their generation would join them each year. However, the dead are now digitally mediated to reside within our everyday technologies, allowing us to communicate and interact with them in the same places that we socialise with the living. There can be little doubt that the next decade offers enormous market opportunities for the Digital Afterlife Industry. However, there are also ethical and moral challenges that need facing. I hope this book will be seen as a 'canary in the coal-mine' for this commercially driven sector by warning of the potential harms these exciting and revolutionary technologies can produce. The writing of this book and the research behind it have not been easy. Although it involved looking at the innovative technologies behind the creation of avatars, thanabots, digital zombies and virtual beings, its main focus is on death, dying and bereavement. Many thanatologists would probably agree that this type of research can be tough on the researcher. However, even though difficult, listening to participants who were willing to share their experiences of digital inheritance was one of the most rewarding experiences of my life; here in this book, I will be sharing their stories.

# Acknowledgements

When I started my PhD research at the University of Warwick, I had no idea that it would lead to the publication of this book. The constant advice and encouragement of others have resulted in this work, and I would especially like to thank a few of them here. My PhD supervisor Steve Fuller planted the seed by speaking about 'the book' from the early years of the research and continues to inspire me with his thoughts and knowledge today. Many of the ideas within Chap. 3 around Digital *Dasein* were developed during meetings and conversations with Steve, and I thank him for sharing his knowledge and for his patience. Also, during my time at Warwick, I was lucky enough to be supervised by Simon Williams whose input kept me on track and moving forward until the very end.

I am so grateful to David Baker, who inspired me during my time at the Open University and opened my eyes to the joy of learning. Thank you for your feedback and advice throughout my PhD and your encouragement during the writing of this book. Sarah Watson, what can I say? Your endless WhatsApp messages, GIFs and texts have prevented this book from being thrown into the digital dustbin many times. If you are reading this message in the printed book, you succeeded making sure I never gave up—thank you.

Coming from a profit-driven commercial background, I am still amazed by the support, time and advice freely given by other academics and colleagues. Special thanks to Linda Kaye, Caroline Lloyd, Elaine Kasket and Maggie Savin-Baden: over the past few years, you have been so inspirational, and your encouragement has kept me going during the difficult phases of research and the writing of this book. Thanks also to Jane

Weaver, the 'knower of all things', your enthusiasm has contributed to the finished product. They say the older you get, the more you need people around who knew you when you were young: sincere thanks and love to my childhood friend Sharon Habib, your guidance and advice has always been filled with laughter and tears. You may be geographically on the other side of this planet, but you are always by my side.

The reviewers of this book gave me much to think about, and I would like to thank them for their time, valuable comments and most of all for their belief that the book was worth writing. For me, writing anything for publication is both mentally and emotionally challenging. So here I would like to show my appreciation to the author Stephen King for writing the short story 'Rat' which is about an author who toiled over each and every word of his novel and, shackled by the rituals involved in writing, made a deal with the devil in order to get the book finished—you saved me from thinking it was just me.

Without my family's support and encouragement, I would not have started, let alone completed my PhD, or this book, thank you all for your never-ending belief in me. Finally, I would like to acknowledge the time and emotional energy given by all those who took part in this research—thank you for sharing your darkest moments, deepest grief and undying love for those you have lost.

# Contents

| | | |
|---|---|---|
| 1 | **Introduction: Contextualising Digital Afterlives** | 1 |
| | *But We've Always Remembered the Dead, Haven't We?* | 2 |
| | *Clarifying the Taxonomy* | 4 |
| | *The Internet as a Site for Memorialisation* | 7 |
| | *Digital Immortality and the Fourth Path* | 11 |
| | *Discussing Death: From Taboo to Cyber Space* | 12 |
| | *The Lived Experience of Dying* | 13 |
| | *The Nitty Gritty* | 15 |
| | *The Radio Host and the Tweet* | 17 |
| | *Exploring Digital Afterlives: An Overview of the Chapters* | 18 |
| | *References* | 22 |
| 2 | **The Service Providers: Both Intentional and Accidental** | 27 |
| | *It's Personal: The Motivations of the SPs* | 28 |
| | *Intentional Digital Afterlife Platforms* | 30 |
| | *Messages from The Dead: One-Way Digital Afterlife Platforms* | 33 |
| | *The Reanimation of the Dead: Two-Way Digital Afterlife Service Providers* | 36 |
| | *Accidental Digital Afterlife Platforms* | 41 |
| | *References* | 45 |

**3 A Philosophical Detour** 47
*The Digital Resurrection of the Dead and Socially Active Digital Zombies* 48
*The Changing Landscape of the Uncanny Valley* 50
*Real or Virtual?* 51
*Where the Digital Dead Reside* 55
*A Digital* Nachlass 56
*References* 58

**4 From Digital Footprints to the Ultimate Selfie: The Experiences and Motivations of Digital Creators** 61
*Research Lurking: The Public/Private Dichotomy of Social Media Platforms* 62
*Digital Scholarship and Digital Sociology* 64
*The Motivations of Digital Creators* 65
*Benevolence* 66
*Therapy* 72
*Continuing Bonds* 74
*Technophilia* 77
*The Quest for Digital Immortality* 79
*References* 84

**5 Why Do Digital Afterlives Matter? The Experiences and Motivations of Digital Inheritors** 87
*Collecting Data: Using Skype, the Why and How* 88
*Conducting Qualitative Research with the Bereaved* 90
*Introducing the DI Participants* 91
*The Link Between Access and Control* 93
*The Voices of the Dead* 95
*Timing, Frequency and Triggers* 99
*Timed Posthumous Messages: The Unintended Consequences* 107
*The Digital Versus the Physical* 109
*The Essence of the Dead* 111
*The Comfort, Disruption and Confusion of Digital Afterlives* 112
*The Comfort of Knowing Others Care* 116
*Grief in the Era of COVID-19* 117
*References* 119

**6 Losing the Data of the Dead and Expanding Existing Models of Bereavement** 123
*Sociology and Death* 123
*The Fear of Second Loss* 125
*Grief and Bereavement: The Theories* 132
*Stage Theory* 132
*The Dual Process Model* 132
*Continuing Bonds Theory* 134
*The Internet and Continuing Bonds Theory* 135
*Off the Beaten Track: Growing Around Grief* 136
*Digital Dislocation: The Disturbance to the Dual Process Model* 137
*Digital Bonds: An Expansion of Continuing Bonds Theory* 140
*References* 143

**7 The Future of Digital Death** 147
*Designing for Mourning in a Digital Society* 148
*Designing* Without *Thanatosensitive Consideration* 149
*Just Because You Can …* 150
*Remembering and Forgetting the Digital Dead* 153
*DDNR: Digital Do Not Reanimate* 156
*Profit and Loss: A Voluntary Code of Conduct for the Digital Afterlife Industry* 158
*References* 162

**8 Final Thoughts and Reflection** 167
*Self-reflection: Honesty and Learning* 167
*#saytheirname* 170
*When Participants 'out' Themselves* 172
*The Limitations* 174
*A Final Word* 175
*Further Resources* 177
*References* 185

**Index** 189

# About the Author

**Debra Bassett** obtained her PhD from the University of Warwick. Debra's qualitative research focuses on how through avatar creation, blogs, vlogs and social network sites, people are creating and nurturing bonds with the dead, and how this human-computer interaction and the digital endurance it facilitates may affect how people grieve. She is a Visiting Fellow at the Centre for Death and Society (CDAS) at the University of Bath.

# Abbreviations

| | |
|---|---|
| AI | Artificial intelligence |
| AVAs | Ancestor Veneration Avatars |
| DAI | Digital afterlife industry |
| DC | Digital Creator |
| DDNR | Digital do not reanimate |
| DI | Digital Inheritor |
| DMAM | Digital memories and messages |
| DNR | Do not resuscitate |
| SNS | Social network sites |
| SP | Service Provider |
| STS | Science Technology and Society |
| TSD | Thanatosensitive design |
| UK | The United Kingdom |
| USA | The United States of America |

# List of Figures


| | | |
|---|---|---|
| Fig. 6.1 | The Dual Process Model of Coping with Bereavement. (Schut and Stroebe 1999) | 133 |
| Fig. 6.2 | Expanded model of the Dual Process Model of bereavement | 139 |
| Fig. 7.1 | Remembering and forgetting the dead | 153 |
| Fig. 7.2 | Stakeholders involved in the creation and inheritance of digital memories and messages | 159 |

# List of Tables


| | | |
|---|---|---|
| Table 1.1 | Categorisation of participants involved in the creation of digital afterlives | 6 |
| Table 1.2 | Categorisation of the intentionality of the DCs | 6 |
| Table 1.3 | Categorisation of the intentionality of the SPs | 6 |
| Table 5.1 | Details of the 23 DI participants | 92 |
| Table 7.1 | Sample control menu | 153 |

CHAPTER 1

# Introduction: Contextualising Digital Afterlives

*Did you know that a man is not dead while his name is still spoken?* (Going Postal – Terry Pratchett)

We are all going to die, but in the words of Freddie Mercury 'Who Wants to Live Forever?' We live in a digital era, where social media is a part of the everyday lives of many. Social media platforms were designed for the living; however, they are being used to nurture ongoing relationships with the dead and are increasingly being used to discuss death, dying and grieving. In this digitised world, technology exists which enables us to create avatars allowing us to 'live forever' and advise future generations as reanimated digital zombies. The convergence of death-related issues and technology has become a growing and important area of study across many disciplines, including Sociology; Human-Computer Interaction (HCI); Science, Technology and Society (STS); Cyberpsychology; Death Studies; and Psychology. In our digital society, ubiquitous smart technology ensures the dead permeate into the everyday lives of the living; as sometimes they wait patiently in a state of suspension for the swipe of a finger or the click of a mouse to conjure them back into existence. Adopting qualitative methods and a constructivist grounded theory methodology, this study explores the nature and impact of the creation and inheritance of digital afterlives.



D. J. Bassett, *The Creation and Inheritance of Digital Afterlives*, Palgrave Studies in the Future of Humanity and its Successors, https://doi.org/10.1007/978-3-030-91684-8_1

By using the voices of research participants, in this book I aim to demonstrate how, for the bereaved, digital afterlives are far more than ones and zeros—because for many who inherit them, they are perceived to contain the *essence* of the dead. This holistic study is unique as it presents findings from three distinct groups: owners and managers of the platforms that enable digital afterlives, those creating digital afterlives and those who inherit digital afterlives following the death of the user. The technologically mediated dead are more socially active than at any time in human history and protecting the rights of these socially active dead is crucial to any study of this kind. However, throughout this book, I will be arguing that digital afterlife service providers should consider giving priority to the living; because giving those who inherit digital memories and messages control over their timings and access could play a crucial part in how these precious digital possessions are experienced by the bereaved.

## But We've Always Remembered the Dead, Haven't We?

In this book, I am claiming memories and messages created and inherited using digital technology have changed our relationships with the dead. Moreover, I argue that for many, these posthumous digital artefacts contain the *essence* of the dead in a way that physical keepsakes do not. In the none-too-distant past, the dead remained physically separate from the living; buried and visited in graveyards; remembered by the lighting of a candle, or glancing through cherished photographs. But these acts of remembrance are deliberate: we can control their timings and frequency. However, today in our digital society—and in ever-growing numbers—we talk to the dead not through thought and prayer but through WhatsApp, Twitter, Facebook or other social networks as we carry the dead with us on our everyday digital devices. We can store the dead in the cloud and hold them in our pockets as they sit there in a state of suspension, until we conjure them back to life with the swipe of a finger or, importantly, when algorithms dictate.

The proliferation of social media platforms and the omnipresent nature of smart technologies have resulted in social media becoming a fundamental part of people's lives in the early twenty-first century. These social networking sites have become part of the everyday 'norms' and routines of many people, and throughout this book, I will examine how the ubiquitous nature of smartphones and wearable technologies is not only changing our day-to-day lives, but also changing the way people grieve; and crucially the way people discuss death and dying.

Since its advent in 1982, the Internet has changed our cultural understanding of social interactions and personal relationships by mediating our daily lives. However, while some use social networking platforms to report important—and not so important—life events, others use these sites to discuss the profound, complex and emotive issues of death, dying and grieving. Before the invention of the Internet and the digital era it gave birth to, how we remembered the dead was a matter of individual and collective memory through the stories and biographies we painted of our loved ones, sometimes in abstract, sometimes in fine detail. However, these precious memories and messages are now mediated through apps, bots, small enterprises and global companies. Moreover, this new Internet-enabled griefscape is expanding at pace; it is estimated by 2100, around 1.4 billion Facebook users will have died, an estimate which assumes Facebook still exists in 2100 but attracts no new users (see Öhman and Watson 2019).

This study is based on PhD fieldwork, and through the following chapters, I will explain why digital afterlives are an important area of research which requires ongoing attention and input from a wide variety of professional and academic spheres. This book is an invitation to consider how our social media presence—including digital messages, social network profiles, thanablogs, posthumous chatbots, posthumous messages and posthumous avatars—engender digital afterlives and how the creators and inheritors experience these mediated digital afterlives. Moreover, I hope this book will also serve as a cautionary tale—not only to those providing digital afterlife platforms—but also to those providing guidance and care to the bereaved and the dying. The year 2021 has given impetus to books and research addressing digital death and dying: the restrictions placed on much of the world's population due to the COVID-19 pandemic have forced many people to attend virtual funerals and use apps such as FaceTime to say their final goodbyes to loved ones. Digital afterlives enabled by the Internet are here to stay, and this book explains their importance by presenting empirical qualitative research, providing the first holistic study of this type. The rationale behind the research—underpinned by the theoretical perspective of symbolic interactionism—has close links with constructivist grounded theory and thanatology. Symbolic interactionism assumes individuals are active, creative and reflective, and that through human interaction, meaning is constructed and takes place in a social world (Belgrave and Charmaz 2014). This perspective seems apt when the field of study is social network sites (SNS). This introduction sets the scene for the rest of the book, and it paves the way for the more philosophical aspects surrounding digital messages and memories which are woven throughout the following chapters.

## Clarifying the Taxonomy

While familiarising myself with the existing literature during the early stages of the research, I wrote a paper calling for further discussions around the terminology used by thanatologists exploring posthumous digital endurance (Bassett 2015). Since this area of research was relatively new, no suitable framework existed to categorise what I identified as three distinct areas: the 'users'; the types of messages and memories created; and the types of platforms utilised to create and enable digital afterlives. Therefore, in order to fully explore how creators and inheritors of digital afterlives experience digital artefacts, I devised a framework that I adopted throughout the research. This short section introduces and clarifies these terms and can be used as a reference section when reading the following chapters.

First, I became aware that the term 'digital legacy' was used interchangeably to describe very different aspects of online, posthumous digital artefacts. This interchangeability may seem like a potato/potahto discussion, and maybe we should call the whole thing off. However, it starts to matter when discussing what these things mean to the creators and inheritors of digital afterlives. I was not the only thanatologists thinking about this nascent terminology: Pitsillides, Waller and Fairfax address the issue by including a key terms and definition section in their book chapter. They used the term 'digital heritage' to describe 'The accumulation and curation of digital data online, which could form the basis of an inexhaustible resource containing the exact documentation of our digital past', while the term 'digital historical artefacts' was used to describe 'digital objects, which contain information for the building up of archives or *Digital Heritage*: for example Social Networks' (Pitsillides et al. 2013). However, I still felt this did not go far enough, and the phenomena needed further clarification. I suggested 'digital legacy' should be used to discuss digital data, passwords, account information, digital assets and digital property: in essence, things that belong in a digital safe or vault that are static once the user has died. However, I suggested the term 'digital selves' should be used for memories and messages: data representing our digital selves: personal videos, messages, photographs and blogs, which belong in a 'digital memory box' rather than a 'digital vault'. Since writing this paper in 2015 I have somewhat changed my mind. As we shall see later in Chap. 3, the term 'digital selves' is wrapped up with the philosophical debates around 'persons' and 'selves'; therefore, in this book, I adopt 'digital memories and messages' to describe the items we would keep in digital memory boxes.

In the early stages of the fieldwork, the term 'digital immortality' seemed an appropriate way to describe the research and certainly caught people's attention. However, although the term digital immortality is relatively new, I argue that it is already an outdated term that is inept when used to describe the 'posthumous persistence of digital data' (Kasket, 2012). Just as we are biologically mortal, we are also digitally mortal and the risk of contagion or the loss or deletion of data that can cause digital erasure is a real threat to the notion of achieving digital 'immortality'. Early in the first phase of data collection, I wrote an email explaining the study to a potential participant who replied that they found the term 'digital immortality' insensitive and painful as it insinuated that their loved one had not died. I then considered adopting Kastenbaum's term of 'assisted immortality' (2016), but this still implied a state of non-death. Reflecting further, I decided the term 'digital endurance' was more suitable and changed the details on the information sheet and online biographies. However, later in the fieldwork, I realised that a change of terminology was necessary *throughout* the literature focusing on thanatechnology. Not just to be sensitive to the needs of the bereaved, but as a more practical—and importantly more accurate—way to describe how people experienced digital memories and messages either by creating or by inheriting them.

So, with some of the terminology in place, in 2016, I began recruiting participants. However, this also presented problems as I realised that treating participants as a single group was unhelpful. Their experiences and motivations differed depending on their role in creating, inheriting and enabling digital afterlives. Therefore, it was necessary to distinguish between the various people who were going to be involved in the research, as they had different—but sometimes overlapping—characteristics. Table 1.1 shows the categorisation of the stakeholders involved in digital afterlives. However, it should be noted that these participant categories are not mutually exclusive, as some participants fall into more than one category.

**Table 1.1** Categorisation of participants involved in the creation of digital afterlives

| *Category* | *Abbreviation* | *Description* |
|---|---|---|
| Digital Creator | DC | The creators of posthumously persistent digital memories and messages |
| Digital Inheritor | DI | Those who inherit the digital memories and messages enabled by thanatechnology |
| Service Provider | SP | Company whose websites, SNS and Apps enable the creation of enduring digital artefacts |

**Table 1.2** Categorisation of the intentionality of the DCs

| *Category* | *Description* |
|---|---|
| Intentional digital message or memory | Messages intentionally created to be inherited by DIs<br>OR DIs messaging the dead as though still alive |
| Accidental digital message or memory | Not intentionally created to be delivered posthumously |

**Table 1.3** Categorisation of the intentionality of the SPs

| *Category* | *Purpose* | *Example* |
|---|---|---|
| Accidental posthumous site | Designed for the living | Facebook, Twitter |
| Intentional posthumous site | Designed for the dead | LifeNaut, Eternime |

I also noted that all types of posthumous digital memories and messages were discussed as if they were one category. I identified a need to distinguish between messages and memories *intentionally* created to be left after creator's death and those that *accidentally* become posthumous digital artefacts. Table 1.2 shows this vital distinction.

Following this distinction, and once again adopting 'intentionality' as a basis, I categorised the two different types of platforms involved in creating and storing posthumous digital memories and messages: Table 1.3 shows this further distinction. Interestingly, as we will see, the accidental SPs seem more willing—or perhaps better funded—to react to some of the needs of the DIs. Although, as this book will argue, more can and should be done to enable DIs to control certain aspects of the digital memories and messages they inherit.

Whilst I am delighted other academics have adopted this taxonomy (Clabburn 2018; Fuller 2019; Matthee 2019; Savin-Baden and Burden 2019; Savin-Baden and Mason-Robbie 2020; Sofka 2020), I hope to spark further discussions between academics and professionals by re-presenting it here.

Thanatologists widely accept that the Internet is changing the way some people grieve (Brubaker et al. 2013; Sofka et al. 2012; Walter et al. 2012). Humanity is becoming ever more digitised (Fuller 2011; Lupton 2015), therefore it is easy to understand that this digitisation of life and society would change how we remember and memorialise the dead. In his book *The Revival of Death*, Tony Walter suggests dying, funeral practices, mourning and beliefs about the afterlife are continually evolving (Walter 1994). Throughout this book I suggest that technology is accelerating this evolution exponentially, and that, as a result of this accelerated evolution, some practitioners are struggling to keep pace with the way people are using the Internet to grieve. Giving further context to the rest of the book, in the next section, I present a potted history of the thanatological research looking at how the Internet has become a fundamental and essential tool for the bereaved who live in digitally connected societies of the West.

## The Internet as a Site for Memorialisation

In 2006, Facebook was launched to anyone with an email account as a social networking website where people could create an online profile and connect to others. However, in 2007, Facebook dealt with the death of some of its users following the Virginia Tech shootings. At that time relatives requested that the Facebook pages of those killed—mainly young students—be left active. Facebook decided to memorialise the accounts rather than switch them off, enabling friends and relatives to write messages of support in their time of grief. The immediacy of Facebook, along with online obituary pages and memorial websites such as Legacy.com, enables people to use social media for public grieving. From the first cave drawings to the magnificent Egyptian mausoleums, when we look back in history, we see symbols of how humans have attempted to leave their mark on the world and, in doing so, have been remembered for millennia (Sofka 1997). Generally, mausoleums are the preserve of the rich and famous, and historically, clergy, obituary writers and eulogy writers had the job of providing the narratives of the 'common' dead. However, for the digitally connected, the Internet enables millions of people to embark on a quest to be remembered by leaving digital traces that can be stored and inherited. How long these digital memories are stored and retrievable is the primary concern of this book.

The vast amount of personal information and the auto-biographical nature of the data posted on social network sites (SNS) provide a rich narrative of people's lives for others to draw upon following the death of the user. Ellison describes SNS as 'web based services that allow individuals to (1) construct a public or semi-public profile within a bounded system, (2) articulate a list of other users with who they share a connection, and, (3) view and traverse their list of connections and those made by others within the system' (Ellison 2007). They provide a mediated environment for communication in a networked age and quickly became a new site for grieving (Brubaker et al. 2013). In their 2013 paper, 'Beyond the Grave: Facebook as a Site for the Expansion of Death and Mourning', Brubaker et al. consider how SNS produce new arenas for grieving, and they explain how 'Facebook is associated with an expansion of death-related experiences—temporally, spatially, and socially'. They argue that these arenas create a socially situated space where the bereaved can engage. Moreover, they suggest these new spaces are situated in people's daily lives, thus ensuring death becomes part of the everyday lives of many. However, as Walter et al. (2012) argue, rather than just making grief public, SNS enable grief to 'become more communal'.

The Internet and the proliferation of smart technologies change the landscape of grief from the physical to the virtual (I will pick up this important point later in this book). In her seminal paper, Sofka (1997) coined the term 'thanatechnology' to describe the convergence of death and technology. In their 2012 book, editors Sofka, Cupit and Gilbert provide a detailed overview of how dynamic and fast-moving social media platforms, such as Twitter, Facebook Skype and YouTube, provide challenges and opportunities for diverse groups such as counsellors, researchers and death educators. The authors argue that these platforms allow for the datafication of end-of-life communications and the creation of 'new sites' of grieving (Brubaker et al. 2013; Walter et al. 2012). It is the impact of thanatechnology, in all its guises, that forms the basis of this book. And as will be discussed later, WhatsApp and LinkedIn—which are absent from the list of platforms listed by Sofka et al.—are mentioned by the bereaved as favoured platforms for creating, storing and retrieving digital memories and messages.

The ramification of thanatechnology and the phenomena it produces has resulted in a growing body of research and literature. Much of the research has focused on how the bereaved are using SNS specifically to discuss their grief. In 2013, Brubaker et al. conducted an inductive analysis of interviews to examine how online identities persist after a user's death and how friends and family use these sites as a space to grieve. It concludes that rather than causing a disruption to the traditional practices and rituals of grieving and memorialisation, SNS provide a 'socially situated' space to grieve which is a part of the daily lives of the bereaved (Brubaker et al. 2013). The work completed by Brubaker et al. focuses specifically on Facebook; however, the research presented in this book examines whether this is the case for other SNS.

Other researchers are exploring these new sites for grieving: Carroll and Landry's research suggests that those marginalised by traditional forms of grief use SNS to stay connected to the dead. These sites enable and empower groups of disenfranchised grievers to feel like part of a grieving community (Carroll and Landry 2010). Doka defined disenfranchised grief as 'grief experienced by people that is not or cannot be openly acknowledged, publicly mourned or socially supported' (Doka 1999). SNS enable the creation of support groups and memorial sites for those bereaved following aids-related death, suicide loss, pet bereavement or perinatal loss (Borgstrom 2017; Sofka 1997; Walter et al. 2012).

The public nature of grieving on the Internet has both positive and negative consequences: Marwick and Ellison's 2012 paper discusses how SNS can be problematic for those engaged in impression management of the dead (attempting to influence how others perceive the dead) and how context collapse happens when diverse social categories (e.g. work colleagues, acquaintances, family, friends, close friends, school peer groups) collapse into categories of 'friends' or 'contacts' resulting in the flattening of traditional notions of hierarchies (Marwick and Ellison 2012). When 'grief tourists' and Internet trolls get involved in communications regarding the dead on SNS, impression management by those who inherit the data of the dead can be compromised (Marwick and Ellison 2012). Other researchers have examined the complexities of online trolls and grief tourists who target Facebook memorial pages and the impact that these RIP trolls have on the bereaved (Phillips 2011; Riechers 2012;

Seigfried-Spellar and Chowdhury 2017). In a 2012 paper 'Does the Internet Change How We Die and Mourn? An Overview', Walter et al. were the first to summarise the possible implications of this fast and dynamic phenomena on thanatology. They suggested there are three types of death: traditional, modern and postmodern (Walter et al. 2012). Traditional death took place in the community, which shared a common theology, and thus produced a bereaved community. Modern death happens at home or in a hospital where death becomes private, hidden away, medicalised and sequestrated (Giddens 1991). Thus, this 'modern death' ensures death does not disrupt the everyday routines of the living (Mellor and Shilling 1993). According to Mellor, death disturbs Giddens' notion of 'ontological security' (Mellor 1992). However, Walter et al. (2012) suggest that, in postmodern death, the public and the private are combined: SNS communications produce private feelings expressed in public spaces (Brubaker et al. 2013; Kasket 2012; Marwick and Ellison 2012; Pennington 2013). Furthermore, Walter et al. (2012) argue this then creates a community of 'diverse mourners' supporting the notion that the Internet is changing how people deal with death, dying and grieving.

Those who research online memorials (Bouc et al. 2016; Getty et al. 2011; Harju et al. 2014; Kasket 2012; Marwick and Ellison 2012; Pennington 2013; Valentine 2008) have found that continuing bonds theory is a useful framework to describe how technology is enabling the bereaved to continue a relationship with the deceased. In a 2013 paper, Pennington found that Facebook users 'Don't De-Friend the Dead' (Pennington 2013). Instead, she suggests the living 'renegotiate' their relationship with the dead. This active process and restructuring of the relationship echo Walter's 'durable biography' (Walter 1996) and Neimeyer's 'meaning making' (Neimeyer 2001); however, with SNS, this restructuring happens in a virtual world rather than in the real world. In Chap. 3, we see why treating these two 'worlds' as dichotomous spaces is unhelpful when researching digital afterlives. In her 2013 research, Pennington examined Facebook profiles of college students to analyse grief communications; she found the majority of participants—the 'silent majority'—did not actively engage with the site of a deceased friend, although most found support by 'lurking' on these sites. The 'vocal minority'—those who did post messages—addressed their posts directly to the dead, supporting the theory that some believe that online 'the dead are listening' (Kasket 2012). Pennington also notes how some students hide the posts written by dead friends from their timelines because they find them distressing (Pennington 2013). Brubaker et al. also note that caution is needed, as, for some, public expressions of grief are distressing (Brubaker et al. 2013). Social media may provide a space for a renegotiated relationship with the dead; however, it is important to acknowledge that grief is idiosyncratic and not everyone grieves in the same way.

## Digital Immortality and the Fourth Path

The notion of 'digital immortality' can be seen as a relatively new variation of what Robert Jay Litton describes as symbolic immortality: the need to pass on memories and attempts to influence future generations (Litton 1979). This quest for immortality is certainly not new: in his 2007 text, Robert Kastenbaum illustrates how the promise of an afterlife is woven throughout the world's main religions, and how through history, people have attempted to continue relationships with the dead. From stone monuments and time capsules to cloning and reproducing, people have always pursued immortality in an attempt to escape death or avoid nonexistence; since the invention of technology that enables voice recordings, technology has mediated this 'assisted immortality'—or assisted survival (Kastenbaum 2016).

Technologies available in the digital age give us perhaps the greatest opportunity to explore the possibilities of assisted immortality. However, as we will see later in this book, this may only be available to those who can afford to pay for the storage and retrieval of digital memories and messages. Writing in 2017, Tony Walter suggests the dead survive by (1) becoming ancestors, (2) becoming immortal or (3) surviving in the memories of the living (Walter 2017). Philosopher Stephen Cave suggests four narratives that dominate the path to immortality: (1) Stay alive—the cheating of death with an elixir of life, (2) Resurrection—physically rising from the dead in the body we inhabited in life, (3) Soul—spiritual survival where the body is not needed and (4) Legacy—with everlasting fame (Cave 2012).

Each time we interact online, we generate digital crumbs, creating what is described by Paul-Choudhury as 'digital souls' (Paul-Choudhury 2011). These 'narrative bits' of information are what Mitra calls 'narbs' (Mitra 2010). The more we interact with technology, the more our digital selves evolve into a realistic representation of our 'real' selves. Thanatechnology enables the proliferation of these 'posthumously persistent' (Kasket 2012) narbs and offers the promise of immortality by way of a digital afterlife. Mitra explains:

> *The identity of an individual is eventually constructed by the combination of narbs that are available on a social networking site where different kinds of narbs work together to produce the composite narrative of a person at any moment in time.* (Mitra 2010, p. 12)

Writing about virtual worlds in 2011, Braman, Dudley and Vincenti suggest that the more we inhabit virtual worlds, what we leave behind posthumously becomes important for some to consider and protect (Braman et al. 2011).

## Discussing Death: From Taboo to Cyber Space

Mourning is described as the act of making visible responses to grief through the creation and curation of materials (Small 2001). It would be wise to acknowledge that changes to mourning practices are nothing new; mourning practices are not static; they change as social and cultural rituals change. In 1965 Gorer, an English anthropologist, suggested that secularisation in many Western societies had begun to affect the grieving practices of many, as the religious rituals that guided people through grief and mourning were beginning to lose their significance. Gorer suggested that the post-war influence and resulting mass mourning led to a rejection of death and its practices; earlier, in 1955, he suggested that sex and death had switched positions, and death had become the more taboo subject (Aries 1981; Gorer 1955). However, with 24-hour rolling news programmes projecting scenes of war, famine and terrorist attacks into our homes, any notion that discussions of death and dying are taboo may be outdated. Death scholar Tony Walter suggested that in the 1960s people were able to express themselves more openly: following two world wars, Britain was experiencing a time of peace, and the suggestion that it was good to talk about emotions began to gain momentum. This momentum continued, and Walter believes it was fuelled partly by the very public death of English reality-television personality Jade Goody in 2009, which dominated much of the British media during her illness and following her death. The public nature of her death was unprecedented at this time, because as Walter points out, until this event, the rituals and practicalities surrounding the death and dying of ordinary people normally remained private. By the time of her terminal illness diagnosis, Jade Goody was already a well-known celebrity; however, what about those who become famous *because* they are dying?

## The Lived Experience of Dying

The Internet has created a space that allows non-celebrities to die publicly. Available technologies and platforms, such as Twitter, Facebook and YouTube, allow people to create their own digital narratives by writing, filming and publishing details of their lives—and for some, their deaths. These blogs, tweets and videos form a collection of enduring digital memories—digital footprints or digital fingerprints created and stored on various SNS and platforms on the Internet. Thanablogging—a term coined by Carla Sofka—describes how the terminally ill use the Internet to take control of their finite time by blogging about their experiences of end-of-life care and dying (Sofka et al. 2012). Terminally ill doctor Kate Granger—who died in 2016—used the Internet to blog, tweet and make videos during her illness to share her experience of terminal cancer and in the hope of de-stigmatising discussions around death and dying.

Moreover, throughout her terminal illness, and up to a few hours before she died in 2016, Kate Granger used the Twitter handle #deathbedlive to bring end-of-life care discussions into the public sphere. At the time of her death, Kate Granger had over 48,400 Twitter followers and had tweeted over 12,500 times. Interestingly over four years after her death, Kate Granger still has 41,400 Twitter followers, supporting research by Natalie Pennington who found people 'don't de-friend the dead' (Pennington 2013). Before social media platforms were part of everyday life for many in our digital society, there were other public dying and deathbed fame examples, such as Christopher Hitchens in 2011 (Brennan 2015). However, Jade Goody and Christopher Hitchens were both publicly known before their terminal illnesses, while Kate Granger reached the status of what I term 'thanacelebrity'—someone who becomes famous *because* they are dying. Thanablogging is a digital version of autopathographies, but the immediacy of SNS ensures these messages—from what Hjorth and Moon describe as the 'soon to be dead' (2017)—feel distinctly different. Later we will look at the motivations and experiences of those creating thanablogs, and the experience of those who inherit them following the death of a loved one.

It seems that discussions about death in the public sphere seem to be growing: Death Over Dinner is an organisation that encourages conversations about what people would like to happen when they die—and importantly, while they are dying (Death over Dinner 2017). It hopes to get people discussing important issues about death and dying over the dinner tables of the world. People's wishes in terms of end-of-life care are often complex. Whether they want to die at home or in an institutional setting, what type of funeral they would like and how they want their body disposing of are all issues that the organisation suggests should be discussed openly over the dinner table. Since 'death over dinner' launched in 2013, an estimated 100,000 parties have been held in over 30 countries (Thrive Global 2017). The recent emergence of death cafes also signals a move that more open debates about death and dying are occurring in the public sphere. These death cafés are not altogether new: their roots lie in salons and coffee shops of the seventeenth century (Brennan 2015). Death cafés take place in over 51 countries worldwide, and in Britain, the first death café was held in London in 2011 by John Underwood (who has since died). However, his mission to 'increase awareness of death with a view to helping people make the most of their (finite) lives' continues to grow. During the COVID-19 pandemic, these 'cafés' moved online in order to continue to support people facing end-of-life choices during the crisis.

How contemporary Western societies deal with death, dying and bereavement is fundamentally social. The death of a loved one is a life-changing event and is experienced not only by individuals but also at the macro societal level. For sociologists, death is a social fact, not just a biological fact: grief affects individuals and broader groups significantly and triggers both social and cultural processes. During the research and whilst writing this book, there have been terror attacks and mass shootings in Paris, Orlando, Manchester, London, Las Vegas, New Zealand and Sri Lanka. The online response to these attacks highlights the social context of death and grief and how public displays of grief are changing in the twenty-first century (Walter 2009). Social media have played an essential part in these social demonstrations of mourning and solidarity: the hashtag #JeSuisCharlie was used over 6500 times per minute following the deadly attack on the offices of Charlie Hebdo in 2015 (Whitehead 2015). The platforms also react to these atrocities: Facebook quickly introduced a French flag meme for their users to incorporate on their profile page following the Paris attack, so their users could show solidarity with the country that had been affected. How people deal with grief is complex; it is entwined and embedded with issues around gender, race, class and culture, or to be more specific—within a social context. It is not just social grief that is context-specific; individual or private grief also happens within a social context that is amplified and made increasingly visible by social media.

## The Nitty Gritty

The methods and ethical framework used for this empirical interview-based research, and the issues and contributions, are presented in the relevant chapters. However, here I think it is helpful to outline the particular theoretical assumptions which underpinned the study. Any attempt to understand social research without a clear understanding of the epistemological and ontological assumptions of the researcher is naive, as these assumptions influence the design—and therefore the findings—of any study.

As a sociologist, I believe attempting to understand how people make meaning of the world around them is the cornerstone of sociological research. With this in mind, in the research I adopted a qualitative methodology which allows for the collection of rich data to gain a contextual understanding of how digital afterlives affect people's experiences of grief in digital Western societies. As a researcher, I am interested in experience rather than causal relationships, and I believe quantitative methodologies cannot achieve this as they fail to highlight the richness and depth of how people experience any given phenomena. Qualitative research can reveal how the practices, protocols and social norms of grief and bereavement may be changing in the emergent spaces created by the Internet. This research was underpinned by a social constructivist perspective, characterised by the assumption that reality is subjective and that social actors continually revise phenomena and their meanings. Furthermore, this ontological position assumes that through interaction, reality is socially and culturally constructed (Charmaz 2014).

Because the study focused on the nascent phenomena of digital memories and messages created and inherited on the Internet, I wanted to be open to the possibility of developing new theories; as existing theories around death, dying and bereavement were developed before the existence of social networking platforms. Grounded theory—the iterative approach to analysing qualitative data—has become one of the most widely used frameworks in qualitative data analysis. Moreover, grounded theory has historical links with thanatology. Its developers Glaser and Strauss studied the awareness of dying and conducted one of the first major studies to use grounded theory as a research process (Glaser and Strauss 1966). This classic study did not have theory as its starting point; rather, it developed theory from carrying out and comparing case studies. Their concern was that using other methods of qualitative data analysis, researchers could manipulate facts to 'fit' theory (Glaser and Strauss 1967, p. 29). This systematic inductive method assumes that theory is 'grounded' in the data, rather than empirically testing existing theories (Flick 2018). Initially, I was drawn to the open-minded approach to data that grounded theory offered. Although I thought it could help me to achieve the research aims, it felt somewhat inflexible in its approach—or lack of approach—to reflexivity. Grounded theory did not reflect my belief that in order for research to be meaningful, the role and biography of the researcher—along with the agency of the participant—should be considered and acknowledged. Moreover, I required a methodological framework that highlighted how knowledge is co-produced by the researcher and the research participants.

In 2000 Kathy Charmaz—building on the works of Glaser and Strauss (1967), and Strauss and Corbin (1990)—developed constructivist grounded theory. This nuanced version of grounded theory preserved the practical strategies developed by Glaser and Strauss and Strauss and Corbin, but importantly for my research interests, integrated more recent developments in qualitative research such as reflexivity and the co-construction of data. Constructivist grounded theory draws on the ontological perspective of constructivism, which assumes people—including researchers—construct the realities in which they participate. Moreover, it acknowledges that a researcher's interpretation of a phenomenon is itself a construction (Charmaz 2014).

Belgrave and Charmaz (2014) suggest that constructivist grounded theory can play a significant role in the field of thanatology as it goes deeper into participants' meanings of illness, dying and death. The theoretical perspective of symbolic interaction also provided a cornerstone to the research and is tightly linked to both constructivist grounded theory and thanatology. It assumes individuals are active, creative and reflective (Charmaz 2014, p. 345) and that in and through human interaction, meaning is constructed and takes place in a social world. Building on the work of George Herbert Mead, Blumer coined the term symbolic interactionism to argue that individuals continually interpret the symbolic meaning of their environments—which includes the actions of others—and meaning is constructed through interaction, language and interpretation (Blumer 1962). In other words, it examines human behaviour and the meanings people attach to elements of their lives. Goffman (1959) claimed that people act and react along similar means and constraints universally, and a unified image of reality is achieved by collective understanding. This 'collective understanding' is crucial to thanatological research as any knowledge about grief cannot be culture-free; rather, grief is shaped by sociocultural processes which help to establish the meaning of a loss (Neimeyer 2001; Rosenblatt 2008). Taking a symbolic interactionist approach to the research allowed me to highlight how individuals are influenced by—and in turn influence—the social norms surrounding loss and grief.

## The Radio Host and the Tweet

There is much anecdotal evidence in addition to research data which demonstrates the importance of digital afterlives to those who create or inherit them. However, in February 2018, when I came across a tweet by the UK radio host James O Brien, I began to understand the scale of the phenomena. LBC is a London-based, national talk radio and phone-in station, and O'Brien—who describes himself as a liberal talk show host—is known as an outspoken, no-nonsense combative broadcaster who has clashed with the likes of UK political commentator Nigel Farage. O'Brien tweeted that he had given his daughter his old mobile phone—which had been disconnected from the Internet—to play games on. He learned that his daughter had been texting her grandfather—his father—who had died five years earlier. In one of the messages, she wrote, 'I love you grandad have a great time in heaven I hope you meet Jesus and I hope you had an amazing life I absolutely love you and my life could not of been this amazing with you'. In other messages, she wrote, 'I'm nearly 10 and Elizabeth is 12! She loved your present by the way your present was your love'. After each message, the rather poignant delivery status read 'not delivered'. The tweet was of obvious interest to me and my research into digital afterlives; however, it was the huge response to his tweet that demonstrated the importance of this new phenomena to those following O'Brien on Twitter.

At the time of his tweet, James O'Brien had 405k followers, and this single tweet received 143k likes, 29k retweets and 3.1k replies. O'Brien's Twitter followers expressed how touched they were by the idea that his young daughter was using his old mobile phone in this way. Over the next three days, I monitored the replies to the tweet, and it became apparent that these tweets echoed what I was hearing from the research participants, who were telling me how precious these technologically mediated communications with the dead are. The O'Brien Tweet reaction helped me to further understand the importance and scale of the phenomena I was researching. Below I have presented just a few typical responses to illustrate the emotive way people describe their posthumous digital messages. I have presented the responses in a table rather than as quotations, as I felt it was important to acknowledge that they are single tweets rather than data resulting from the in-depth participant interviews used for the majority of this qualitative research.

*I think this is beautiful. My daughter died in 2009, she was 25, and I messaged her every day for 2 whole years. The messages turned from despair to chats and family news. Sometimes I charge her phone up and ring it just to hear her voice. Not often but once in a while*

*My 11 year old has my dad's old Nokia She texts him all the time, he died 5 years ago makes me so happy sad*

*My grandfather passed away 7 years ago, one day somehow his contact got on to the favourites by chance on my new phone. I text him every bad day. It's therapeutic. Your kid is a good egg mate*

*Beautiful, best thing I have read today. I stumbled upon an answerphone message on my work phone weeks after my father died. Knocked me for six. I saved it and I still have a listen now and again*

I found it compelling that everyone I contacted to ask permission to use their replies to the James O'Brien tweet agreed. The recruitment of both the DI and the DC groups of participants also reflected this agreement and eagerness to talk about experiences of posthumous digital messages and memories. People wanted to discuss the digital afterlives they had inherited following the death of a loved one; moreover, people wanted to talk about the dead.

## Exploring Digital Afterlives: An Overview of the Chapters

My digital ethnographic research took place between October 2015 and September 2019. I choose to take a holistic approach to the topic in an effort to fully explore the effects of digital afterlives. To achieve this I recruited and interviewed people from each of the three distinct groups of shareholders involved in the creation and inheritance of digital memories: service providers (SPs), those who work for the digital afterlife platforms; digital creators (DCs), those who purposefully create digital memories and messages to be left posthumously; and digital inheritors (DIs), those who inherit the digital afterlives following the death of loved ones. Throughout this book, I present empirical research demonstrating how digital afterlives are blurring the distinction between the social lives of the living and the social lives of the dead and the ramifications of this blurring. In the following chapters, I will introduce you to the bereaved, the dying and the people behind the companies who make digital afterlives possible—the participants.

This book is divided into eight chapters. Here in this introductory chapter—using the James O'Brien tweet—I have given just a flavour of how the Internet, and specifically social media platforms, are changing and challenging the way we talk to—and about—the dead. In Chap. 2, I present the data from the SP group of participants—the owners and managers of the platforms that enable the creation of digital afterlives. We will begin by looking at the motivation of these participants before looking at the rate of failure of the platforms *intentionally* offering 'immortality'. Also, in Chap. 2, I will present data that demonstrates the lack of understanding—on the part of many of the intentional SPs—to the potential harms, needs and wishes of those who inherit digital afterlives. As we have seen in this introductory chapter, there are two types of digital afterlife platforms, and the *accidental* platforms are not without their problems, therefore throughout Chap. 2, we will look at issues surrounding data control, RIP trolling and cyberharassment.

In an attempt to provoke further philosophical discussions, in Chap. 3, I put forward the notion that the dead exist in a digital *Dasein*. Thanatechnology enables the dead to exist 'somewhere', and my suggestion of a digital *Dasein* describes this posthumous essentialism, and, as the data presented in this study shows, the idea that the dead continue to be 'somewhere' can bring comfort to the bereaved. This concept of digital *Dasein* is helpful as it highlights the problematic tension between the bounded nature of *Dasein* by reflecting on who's digital *Dasein* we are talking about—those doing the creating or those that inherit digital afterlives. Throughout this chapter, the term 'digital zombies' (Bassett 2015) will describe the posthumous, digital representations of the deceased—whereby the deceased are resurrected and reanimated to do things in death they did not do when alive. Again, focusing on the philosophical nature of digital afterlives, in this chapter I will be contemplating how decisions around the reanimation of the dead create a 'Schrodinger's cat' moment in the digital world. And why, with this in mind, introducing a digital do not reanimate (DNR) order could be a useful addition to intentional digital afterlife platforms.

Digital sociologists and digital scholars will undoubtedly understand problems around Internet lurking and the need to consider the public/private nature of SNS messages carefully. Therefore, in Chap. 4, I address some of the issues I faced with recruitment and data collection. Moving on, this chapter then focuses on those doing the creating: the thanabloggers, creators of posthumous messages, grieving digital creators and the early adopters. Some of the participants created digital memories and messages *knowing* that their creations may live on in digital perpetuity. Through the presentation of empirical data, I hope you will gain a real insight into the hopes and motivations of these digital afterlife creators.

Continuing with the presentation of qualitative data, Chap. 5 focuses on why all of this matters. I start this chapter by clarifying the decision to use Skype for interviewing bereaved participants before introducing you to the digital inheritors, who, by sharing their stories, allow us to begin to understand the importance of digital afterlives and the crucial link between comfort and control. As you will see, the decisions around when and how often to visit the technologically mediated dead can be difficult to negotiate. Therefore, the potential ramifications of receiving timed posthumous messages—pre-record to be delivered on future dates—form an essential part of this chapter. As we will see, the presence of digital afterlives is not a denial of death on the part of the bereaved—they know the dead are dead—however, it appears that for many, the dead are still part of their everyday lives, in ways that they were not before the digital age. Digital afterlives are here to stay, and by the end of this chapter, I hope you will see why it is crucial that all platforms—both accidental and intentional—consider the needs of those who inherit them.

Chapter 6 starts with a potted history of the links between sociology and death to contextualise digital afterlives within the wider academic disciplines. It then considers what happens when inheritors lose the data of the dead: the theory of the fear of second loss—presented in this chapter—describes the anticipatory anxiety caused by the fear of loss, corruption or lack of access to precious posthumous digital artefacts. This chapter then examines existing models of bereavement before suggesting ways they could be expanded to accommodate digital inheritance.

Throughout this book, I will argue that the dead's digital artefacts are experienced by the bereaved differently from physical artefacts. This difference is because, for many inheritors, these digital memories and messages embody the essence of the dead. Therefore, understanding the impact of these enduring digital afterlives and the data of the dead will be crucial in this emerging area of study. Chapter 8 looks at how by adopting Massimi's (2012) notion of thanatosensitive design (death sensitivity), the Digital Afterlife Industry (DAI) could mitigate the problems caused by digital inheritance, which are presented in this book. Just because technologies may exist, which enables the digital reanimation and resurrection of the dead, does not mean they should be used without careful consideration of the inheritors of these digital zombies (Bassett 2015). With this in mind, Chap. 8 concludes with a call for a voluntary code of conduct for the Digital Afterlife Industry (DAI). I suggest this code of conduct could—and should—give inheritors control over the timing of pre-recorded posthumous messages. These commercially driven companies may benefit from implementing such a code, as it could encourage more DCs to use companies committed to strict storage and retrieval policies and who prioritise the needs of those who eventually inherit the digital afterlives they have created.

In the final chapter, I reflect on the effect PhD study and the writing this book has had on me, the researcher, and how my life experiences have fundamentally influenced the fieldwork, data collection and therefore, the findings of this study. This chapter then suggests that changing the names of the dead in qualitative thanatological research is not always the correct thing to do. I will argue that bereaved participants are co-creators of this type of knowledge, and therefore deserve to have a voice when considering whether to change the names of their dead loved ones. This book calls for further research and debate into this phenomenon across multiple scholarly disciplines; however, we need to move faster than—or at least keep pace with—the thanatechnologies being developed and adopted in our digital societies.

Here, at the end of this introductory chapter, I want to be clear about my academic background: I am not a grief councillor, and I am certainly no philosopher. Although much of this book is aimed at these fields of study, I hope that a non-academic audience will find this book of interest. While I hope you will read it cover to cover, I have tried to write the book in a way that the various stakeholders—or those just interested in digital afterlives—will be able to dip in and out of the chapters. In writing this book, I am not claiming to offer a complete understanding of digital afterlives—quite the opposite—we are at the beginning of our understanding of this nascent phenomenon and its impacts. Therefore, I aim to present the research in such a way as to provoke further conversation and debate while at the same time offering some practical suggestions which may be useful to the reader. It is my hope that the following chapters will also provide a cautionary reminder that we live in a digital society, and every time we interact on social media platforms, we are creating and curating digital afterlives which will almost certainly outlive our mortal lives. These digitally mediated messages and memories may indeed endure, and some may live on in perpetuity; or at least until our digital essence disappears and a Google search states we are code 404 file not found.

## References


Aries, P. 1981. *The Hour of Our Death*. New York: Oxford University Press.

Bassett, Debra. 2015. Who Wants to Live Forever? Living, Dying and Grieving in Our Digital Society. *Social Sciences* 4 (4): 1127–1139. https://doi.org/10.3390/socsci4041127.

Belgrave, L., and K. Charmaz. 2014. Studying Illness and Dying Through Constructivist Grounded Theory. In *The Social Construction of Death: Interdisciplinary Perspectives*, ed. L. Van Brussel and N. Carpentier, 34–48. Springer.

Blumer, Herbert. 1962. Society as Symbolic Interaction: 131–139.

Borgstrom, E. 2017. Social Death. *QJM* 110 (1): 5–7. https://doi.org/10.1093/qjmed/hcw183.

Bouc, Amanda, S. Hye Han, and Natalie Pennington. 2016. 'Why Are They Commenting on His Page?': Using Facebook Profile Pages to Continue Connections with the Deceased. *Computers in Human Behavior* 62: 635–643. https://doi.org/10.1016/j.chb.2016.04.027.

Braman, J., A. Dudley, and G. Vincenti. 2011. Death, Social Networks and Virtual Worlds: A Look into the Digital Afterlife. In *Proceedings —2011 9th International Conference on Software Engineering Research, Management and Applications, SERA 2011*, 186–192. Department of Computer and Information Sciences, Towson University. https://doi.org/10.1109/SERA.2011.35.

Brennan, Michael. 2015. Christopher Hitchens' Public Dying: Toward a Secular-Humanist Ars Moriendi: 1–34. https://doi.org/10.1177/0030222815606453.

Brubaker, Jed R., Gillian R. Hayes, and Paul Dourish. 2013. Beyond the Grave: Facebook as a Site for the Expansion of Death and Mourning. *The Information Society* 29 (3): 152–163. https://doi.org/10.1080/01972243.2013.777300.

Carroll, Brian, and Katie Landry. 2010. Logging on and Letting out: Using Online Social Networks to Grieve and to Mourn. *Bulletin of Science, Technology & Society* 30 (5): 341–349.

Cave, Stephen. 2012. *Immortality: The Quest to Live Forever and How It Drives Civilization*. Crown.

Charmaz, K. 2000. Grounded Theory: Objectivist and Contructivist Methods. In *The Handbook of Qualitative Research*, ed. N.K. Denzin and Y. Lincoln. Thousand Oaks: Sage.

Charmaz, Kathy. 2014. *Constructing Grounded Theory*. Sage.

Clabburn, Oliver. 2018. Investigating the Use of Digital Legacies with People Affected by Motor Neurone Disease (MND): An Interpretative Phenomenological Analysis.

Death over Dinner. 2017. Death Over Dinner. 2017. http://deathoverdinner.org/about

Doka, Kenneth J. 1999. Disenfranchised Grief. *Bereavement Care* 18 (3): 37–39.

Elias, Norbert. 2001. *Loneliness of the Dying*. Bloomsbury Publishing.

Ellison, Nicole B. 2007. Social Network Sites: Definition, History, and Scholarship. *Journal of Computer-Mediated Communication* 13 (1): 210–230.

Flick, U. 2018. *Doing Grounded Theory*, ed. U. Flick. London: Sage.

Fuller, S. 2011. *Humanity 2.0*. Basingstoke: Palgrave Macmillan.

Fuller, Steve. 2019. *Nietzschean Meditations: Untimely Thoughts at the Dawn of the Transhuman Era*. Vol. 1. mbassador GmbH.

Getty, Emily, Jessica Cobb, Meryl Gabeler, Christine Nelson, Ellis Weng, and J.T. Hancock. 2011. I Said Your Name in an Empty Room: Grieving and Continuing Bonds on Facebook. In *Proceedings of the 2011 Annual Conference on Human Factors in Computing Systems (CHI '11)*, 997–1000. https://doi.org/10.1145/1978942.1979091.

Giddens, A. 1991. *Modernity and Self-Identity*. 2015th ed. Cambridge: Polity Press.

Glaser, Barney G., and Anselm L. Strauss. 1966. *Awareness of Dying*. Godin: Aldine Transaction.

———. 1967. *The Discovery of Grounded Theory: Strategies for Qualitative Theory*. New Brunswick: Aldine Transaction.

Goffman, Erving. 1959. *The Presentation of Self in Everyday Life*. New York.

Gorer, G. 1955. The Pornography of Death. *Encounter* 5: 49–52.

———. 1965. *Death, Grief & Mourning in Contemporary Britain*. London: Cresset Press.

Harju, Anu A., Katrina Doveling, and Vered (Rose) Shavit. 2014. Researching Digital Memorial Culture and Death Online. 21 (1): 76–81.

Hjorth, Larissa, and Jung Moon. 2017. Visual Afterlife: Posthumous Phone Practices. In *Selfie Citizenship*, 119–126. Springer.

Kasket, Elaine. 2012. Continuing Bonds in the Age of Social Networking: Facebook as a Modern-Day Medium. *Bereavement Care* 31 (2): 62–69. https://doi.org/10.1080/02682621.2012.710493.

Kastenbaum, R. 2007. Death Anxiety.

Kastenbaum, R.J. 2016. *Death, Society, and Human Experience*. 11th ed. Abingdon: Routledge.

Litton, Robert Jay. 1979. *The Broken Connection*. New York: Simon K Schuster.

Lupton, D. 2015. *Digital Sociology*. 1st ed. New York: Routledge.

Marwick, Alice, and Nicole B. Ellison. 2012. 'There Isn't Wifi in Heaven!' Negotiating Visibility on Facebook Memorial Pages. *Journal of Broadcasting & Electronic Media* 56 (3): 378–400.

Massimi, Michael. 2012. Thanatosensitively Designed Technologies for Bereavement Support.

Matthee, Nicolas. 2019. Studying Ritual in Contexts of Online Grief and Bereavement: Methodological Considerations. *Ritual in a Digital Society*: 27.

Mellor, Philip A. 1992. Death in High Modernity: The Contemporary Presence and Absence of Death. *The Sociological Review* 40 (S1): 11–30.

Mellor, Philip A., and Chris Shilling. 1993. Modernity, Self-Identity and the Sequestration of Death. *Sociology* 27 (3): 411.

Mitra, Ananda. 2010. Creating a Presence on Social Networks Via Narbs. *Global Media Journal* 9 (16): 22–40.

Neimeyer, Robert A. 2001. *Meaning Reconstruction & the Experience of Loss*. American Psychological Association.

Öhman, Carl, and David Watson. 2019. Are the Dead Taking Over Facebook? A Big Data Approach to the Future of Death Online. *Big Data & Society* 6: 1–13. https://doi.org/10.1177/2053951719842540.

Paul-Choudhury, Sumit. 2011. Digital Legacy: The Fate of Your Online Soul. *New Scientist* 210 (2809): 41–43.

Pennington, Natalie. 2013. You Don't De-Friend the Dead: An Analysis of Grief Communication by College Students Through Facebook Profiles. *Death Studies* 37 (July 2015): 617–635. https://doi.org/10.1080/07481187.2012.673536.

Phillips, Whitney. 2011. LOLing at Tragedy: Facebook Trolls, Memorial Pages and Resistance to Grief Online. 16 (12): 1–14.

Pitsillides, S., M. Waller, and D. Fairfax. 2013. What Role Does Digital Information Play in the Way We Are (Re)Membered. In *Digital Death*, 75–90.

Pratchett, Terry. 2008. *Going Postal: The Hilarious Novel from the Fantastically Funny Terry Pratchett*. Vol. 33. Random House.

Riechers, Angela. 2012. My Mourning Junkie: Wailing at the Facebook Wall. In *Workshop on Memento Mori: Held in Conjunction with ACM SIGCHI 2012 in Austin, Texas*, vol. 6.

Rosenblatt, Paul C. 2008. Grief Across Cultures: A Review and Research Agenda.

Savin-Baden, Maggi, and David Burden. 2019. Digital Immortality and Virtual Humans. *Postdigital Science and Education* 1 (1): 87–103.
Savin-Baden, Maggi, and Victoria Mason-Robbie. 2020. *Digital Afterlife: Death Matters in a Digital Age*. CRC Press.
Seigfried-Spellar, K., and S. Chowdhury. 2017. Death and Lolz- Understanding the Personality Characteristics of RIP Trolls. In *22nd Annual CyberPsychology, CyberTherapy & Social Networking Conference*.
Small, N. 2001. Theories of Grief: A Critical Review. In *Grief, Mourning and Death Rituals*, ed. J. Hockey, J. Katz, and N. Small. Buckingham: Open University Press.
Sofka, Carla J. 1997. Social Support 'Internetworks.' Caskets for Sale, and More: Thanatology and the Information Superhighway. *Death Studies* 21 (6): 553–574. https://doi.org/10.1080/074811897201778.
Sofka, Carla. 2020. The Transition from Life to the Digital Afterlife. In *Digital Afterlife: Death Matters in a Digital Age*, 57.
Sofka, C., I. Cupit, and K. Gilbert. 2012. In *Dying, Death, and Grief in an Online Universe*, ed. Carla J. Sofka, Illene N. Cupit, and Kathleen R. Gilbert. New York: Springer.
Strauss, Anselm, and Juliet M. Corbin. 1990. *Basics of Qualitative Research: Grounded Theory Procedures and Techniques*. Sage.
Thrive Global. 2017. Thrive Global. 2017. https://journal.thriveglobal.com/michael-hebb-death-over-dinner-38ae6553d21b
Valentine, Christine. 2008. *Bereavement Narratives: Continuing Bonds in the Twenty-First Century*. Routledge.
Walter, Tony. 1994. *The Revival of Death*. Taylor & Francis.
———. 1996. A New Model of Grief: Bereavement and Biography.
———. 2009. Jade's Dying Body: The Ultimate Reality Show. *Sociological Research Online* 14 (5). https://doi.org/10.5153/sro.2061.
———. 2017. How the Dead Survive. In *Postmortal Society*, ed. Hviid Jacobsen, 1st ed., 19–39. Oxford: Routledge.
Walter, Tony, Rachid Hourizi, Wendy Moncur, and Stacey Pitsillides. 2012. Does the Internet Change How We Die and Mourn? Overview and Analysis. *Omega: The Journal of Death and Dying* 64 (4): 275–302. https://doi.org/10.2190/OM.64.4.a.
Whitehead, T. 2015. Paris Charlie Hebdo Attack: Je Suis Charlie Hashtag One of Most Popular in Twitter History. Telegraph Online. 2015. https://www.telegraph.co.uk/news/worldnews/europe/france/11336879/Paris-Charlie-Hebdo-attack-Je-Suis-Charlie-hashtag-one-of-most-popular-in-Twitter-history.html

CHAPTER 2

# The Service Providers: Both Intentional and Accidental

*There is no name that will last forever, even immortality itself is temporal.*
(Bangambiki Habyarimana)

In November 2019, Twitter thought it a good idea to delete some of its inactive accounts, however, it did not anticipate the backlash this decision would have on its users who still visited the accounts of their deceased loved ones. In their defence, it is fair to assume that the creators of ubiquitous social media platforms such as Twitter, Facebook, WhatsApp and LinkedIn intended their websites and apps to be used to communicate *to* the living *by* the living. However, as we have seen in Chap. 1, some have adopted these platforms to communicate with the dead and nurture ongoing relationships with their deceased loved ones.

Platforms and technologies exist today 'intentionally' offering the ability to create digital afterlives: it is precisely what the DAI is designed to do. As we shall see in this chapter, thanabots, deep fakes, avatars and the timed delivery of posthumous messages are just some of the services on offer. In 2019—around the same time Twitter was re-assessing its inactive account policies—in South Korea, a documentary production team were filming a bereaved mother being 'reunited' with her deceased daughter. Her daughter had died four years earlier, aged seven. Using a virtual reality (VR) headset, the mother interacted with her digitally reanimated daughter using real-time rendering technology. This VR technology enabled the mother to talk, touch and feel her daughter before finally saying goodbye.



D. J. Bassett, *The Creation and Inheritance of Digital Afterlives*, Palgrave Studies in the Future of Humanity and its Successors,
https://doi.org/10.1007/978-3-030-91684-8_2

It is almost impossible to write a book about companies offering digital afterlives—and the consequence of such technologies—without referring to the Netflix series Black Mirror. When presenting the research to academic and non-academic audiences, I am inevitably asked if I had ever seen Black Mirror. This is hardly surprising as Charlie Brooker and his co-writers have used the theme of digital afterlives in two of their standout episodes, Be Right Back and San Junipero—San Junipero being my personal favourite (for a fascinating and detailed look at Black Mirror, see The Moral Uncanny in Black Mirror by Gibson and Carden 2020). Black Mirror is described as a dystopian look at the world, but companies facilitating digitally mediated afterlives exist today and are often referred to in the same dystopic way, but is this a fair comparison? In this chapter, using a constructivist grounded theory methodology, I present qualitative data which shines a light on the motivations and hopes of the people behind these digital afterlife technologies.

## It's Personal: The Motivations of the SPs

Between 2016 and 2019, I interviewed a total of eight SP participants. The interviews were via either telephone or Skype. Later, in Chap. 5, we will look at the benefits of using Skype to interview bereaved participants. However, the main benefit of using Skype for data collection from this group of participants was that recruitment could be worldwide without time or cost constraints. All participants in this group lived in North America or Europe, and each qualitative interview lasted between 30 minutes and 90 minutes.

All but one of the SP participants told me how their personal experiences of grief and bereavement had been their primary motivation when creating their digital afterlife platform. Colin is the CEO of an intentional digital afterlife platform, and he explained why starting his company was important to him:

> *being affected by my grandmother's fight with Alzheimer's and losing all her memories long before she passed away.* (Colin)

Ron created—and now manages and owns—an intentional digital afterlife platform that offers the delivery of posthumous digital messages. Ron told me how the death of his father had inspired him to start the company.

*there was half a second of my father speaking and that is the only recording I have anywhere of his voice. So that started the idea that, what if you could have a platform where you could capture people's voices and send them on in the future.* (Ron)

Two other SP interviews echoed the personal motivations of Colin and Ron: Graham, who owns a one-way digital afterlife company, told me:

*I guess I had the idea for 10 years, message from heaven since I lost my father to cancer 14 years ago.* (Graham)

Martyn runs a website that offers support, advice and the delivery of posthumous messages.

*I lost my mother to cancer which is why I started the website.* (Martyn)

As we shall see in Chap. 5, the idea of timed posthumous messages is not without problems for the DIs, and whether to listen to the voices of the dead proved to be a challenging and complex issue within this research. However, it was becoming clear that most of the intentional digital afterlife participants had been motivated by personal experiences. Only one participant was not personally motivated. He told me that he was motivated by the available technology that enabled digital afterlives. Moreover, he wanted to see how far the technology used by the company he worked for could contribute to the concept of digitally backing up a person's mind.

All the interview extracts and examples above are from owners or creators of intentional digital afterlife providers. However, I also found that personal experience was also a motivation for some who work for accidental afterlife providers. I interviewed a participant who worked in the Compassion Department at Facebook (an accidental digital afterlife platform). He explained that each request to memorialise a Facebook profile got a 'human response' and that a team within Facebook handled these requests. This team, he told me, was started by a man whose mother had died from breast cancer.

These very personal motivations took me by surprise; I had wrongly hypothesised that the motivations of the SPs would be grounded in the desire to create and use nascent technology—or for purely commercial reasons. Many of the interview extracts from this group of participants also coded under technophilia; however, the dominant theme was personal motivation. The stories from SP participants reflected the findings from the DC and the DI participant groups: the SP participants had suffered emotional bereavements which had fundamentally affected their lives and the choices they made. Throughout these interviews, the common thread was the desire to design platforms to create, store and retrieve digital memories and messages to nurture ongoing relationships with the dead, and crucially to remember the dead. In Chaps. 4 and 5, we will see if these hopes and passions are experienced positively by the other stakeholders, those doing the creating and those doing the inheriting.

## Intentional Digital Afterlife Platforms

Earlier, in the introduction to this book, I discussed how the term 'immortality' was unsuitable when discussing digital afterlives. However, in this section, I will adopt the term once again, as many platforms offering digital afterlives use this term, and crucially, using it highlights the precarious nature of these platforms. While I am aware that many readers of this book will be familiar with the platforms that claim to offer digital immortality, in this next section, I present just a selection to give context to the rest of the book.

Founded in 2006, The Forever Social based in Sweden offers digital immortality to its users, with the promise that 'you' will be able to respond to world events and other people's posts after you die. By 19 January 2021, The Forever Social had just 25 Twitter followers, with its last Twitter post in April 2018; although their Facebook page is still active, their latest post is still over six months old. Its website is still live, and there is a possibility that its official launch has been delayed rather than abandoned. An interesting aspect of this platform is its explicit rejection of religious belief in its version of a digital afterlife. One of its tweets asks, 'Why believe in God when you can just believe in us? Live on after death. No faith required', while in another tweet, they address God directly: 'Move out of the way, God. The Forever Social is Heaven 2.0 and will keep you connected with loved ones, even after death'. The Forever Social uses a 'Forever Engine', which it claims 'uses algorithms and artificial intelligence to learn how you would respond to everything in the future', suggesting this digital zombie (Bassett 2015) would continue to evolve after the death of the creator; this notion of AI evolution is interesting, and this idea of digital zombies with autonomous agency forms part of Chap. 3.

The Forever Social platform also states:

> *only you can control the legacy you – today and when you're gone. This includes setting the date or trigger for your digital burial, blocking people whose lives you want to exit forever, and limiting who can see your digital testament.* (The Forever Social 2018)

The impact of giving complete control to DCs—even when they are dead—is the crux of this book, and it is precisely the type of statements and sentiments shown in the quotation above—and which are prevalent on digital afterlife platforms—which I will be asking SPs to re-consider in Chap. 7.

In 2010, Don Davidson started Intellitar, a digital immortality platform based in Alabama. By 2015, it had ceased trading. It charged $25 a month to users who wanted to create their digital doubles or 'intelligent avatar', which the site claimed could live on after your death. Like other platforms, the technology company created much media interest but failed to achieve the required number of users to ensure its success. It was a similar story for others: Loggacy, a platform created in 2013, offered users the ability to store their 'most precious memories' and 'connect with family and friends so that you are never forgotten'. On 26 July 2017, Loggacy announced that they were closing the platform and encouraging its users to notify them within 30 days if they would like to retain any uploaded data held on their database. Its website is no longer available online. LivesOn was a web-based application that claimed to be able to continue tweeting for you after you die. Its slogan 'When your heart stops beating you'll keep tweeting' caused much media interest during its inception in 2013. However, continuous monitoring of its Twitter account and website during the research failed to discover any further development in the platform. It has not published any tweets since 5 October 2013, and its website is no longer available.

When considering these intentional digital afterlife platforms, the work of Bell and Gray (2000) is useful. They further categorise the platforms into what they term 'one-way' immortality providers: where the dead are passive (they remain dead) but still preserve and transmit their ideas; and platforms providing 'two-way' immortality: where the dead are active and become reanimated 'digital zombies' (Bassett 2015). I started recruiting participants for the intentional SP group in November 2015. I produced a list of one-way digital immortality platforms that offer users the ability to create digital messages and memories to be delivered posthumously, along with platforms offering two-way immortality where the dead can remain socially active.

Since November 2015—when I first produced the lists—I periodically monitored the status of the platforms up to the end of May 2021. At the beginning of the research, I listed 22 platforms offering digital afterlife services; however, by May 2021, 13 of these platforms either ceased to exist or had no online activity for 12 months or more. This 59% platform failure rate highlights the unstable nature of the 'immortality' industry, and when platforms offering digital immortality or digital afterlives cease to exist, it is crucial to explore what happens to the data they hold. Many of these digital afterlife companies encourage their users to upload precious files with eternal storage and retrieval promises. However, other digital afterlife platforms are unclear about their storage and retrieval policies. During an interview with an intentional SP, Ron explained what his platform provides in terms of longevity of storage, suggesting that 'forever' may only last for 20 years:

> *As we get funds in we're setting them aside so we have funds to pay it ahead if we ever need to. I don't see any reason at the moment why we should do that … but we may at some point see that we can pay ahead for 20 years.* (Ron)

The iterative nature of constructivist grounded theory methods—and how the interview schedule was organised—resulted in simultaneous interviewing and coding transcripts from all three participant groups. During the SP interviews, I was already aware of how posthumous digital memories and messages were being experienced by the bereaved (these findings are presented in Chap. 5). Moreover, I began to understand how platform failure—and therefore data storage and retrieval failure—would result in losing the data of the dead and have profound implications for the bereaved.

The demise of so many platforms offering the storage and the retrieval of posthumous digital memories and messages during the relatively short duration of this research demonstrates the precarious nature of the DAI and its inability to offer any guarantee that the messages and memories stored on the platform will endure. However, as some digital afterlife platforms die, new ones are being created. I suggest exponential growth in the death-tech industry due to technological advances will result in users becoming more accustomed to creating and continuing relationships with the online dead.

In the next section of this chapter, we look at platforms offering one-way 'immortality' services that enable recorded messages to be delivered following the death of their DC.

## Messages from The Dead: One-Way Digital Afterlife Platforms

*May you live to be 100 and may the last voice you hear be mine.* (Frank Sinatra)

Some DCs record memories and messages intentionally to be inherited by others after their death. The impact of such timed messages—or messages where a DI has no control in the delivery of such messages—forms an ongoing thread throughout this book, as we will see in the following chapters. Digital afterlife platforms that work in a one-way direction provide a service whereby the dead remain 'truly' dead: that is, they are not reanimated, socially active, digital zombies (Bassett 2015), or what Nansen et al. describe as the 'restless dead' (2014). Within these one-way digital afterlife networks, the dead are in a static realm. Here in this section, I look at some of the platforms which offer services to DCs by enabling the creation of digital messages to be delivered posthumously and then by notifying people that they have a message waiting for them following the death of the creator.

Started by James Norris in 2012, the UK-based organisation DeadSocial offered an 'end user' planning tool and service, enabling the creation of a 'digital legacy'. Working with hospices, charities and healthcare workers, DeadSocial's user-friendly website aimed to empower those facing end-of-life decisions by enabling the creation and future delivery of messages—including digital memories—to those left behind, as well as practical advice regarding digital legacy issues, such as how to download social media content when someone has died. Analysing the website in 2016, I noticed the prevalent use of religious and spiritual imagery on the site: its logo includes an angel with wings, the title page included a long winding road ascending up to the sky. In its original version, DeadSocial offered a 'goodbye tool' where registered users could 'build a legacy' to be delivered to named inheritors following their deaths. The website contained a link to a tutorial video hosted by YouTube, which informed users that: 'Your first messages is arguably the most important message you will ever leave'. It continued, 'In this message we recommend that you tell your friends and family that you have died'. DeadSocial required you to assign at least one digital executor to control the activation of messages following the death of the DC. The platform enabled the DC to schedule messages to be delivered to others on specific future dates. The ability to nominate a digital executor goes some way to protect the bereaved—which this book advocates—however, it fell short of the control settings I present in Chap. 7. In 2019, DeadSocial was rebranded as MyWishes with a new website which attempts to empower the DC control their online data and prepare for their digital afterlives. Interestingly the newly developed informational and professional website no longer contains religious imagery.

Another company, Safe Beyond, offered the delivery of posthumous digital messages. This paid for posthumous message creation site proposes three types of service (1) dated messages to be released on specific dates; (2) location messages to be released when the DI reached a specific geographic location and, (3) event messages to be delivered on the occurrence of a predefined event. The site's emotional promotional video is a professionally produced video showing a bride receiving such a message on her wedding day. I have used this video at three academic conferences to illustrate the research, and each time it results in gasps from the audience. Wedding rituals change over time—and more importantly, what constitutes a family changes—yet there are many assumptions within this video of a father's posthumous message to his daughter, including the assumption that his daughter would like a message from her dead father on her wedding day. Also, the message assumes that his daughter would marry the 'man' of her dreams and that his wife would not have remarried. The message focuses on a patriarchal social system that sees his son become the head of the family following his death. However, this video is useful to thanatological research as it helps to shine a light on the cultural complexities involved in the creation, and, importantly, the inheritance of posthumous digital messages.

In an effort to protect the messages, SafeBeyond does allow for a trustee to be nominated who would be able to control message delivery, and a tick box provides DCs with the option to reveal or prevent DIs from knowing about the existence of each message. Significantly, however, as with the service offered by MyWishes, this still does not give complete control to the recipient. For example, in their Terms of Service, SafeBeyond explained what happened if their fees were not paid—'If charges are not paid such account may be cancelled and all Digital Assets may be deleted'. The terms and conditions go further—'You agree that any information you provide to the Site and/or Application may be lost or destroyed, and that SafeBeyond is not responsible for the loss of such information'. Once again, it seems that digital afterlives are only available for as long as payments arrive, or the platform does not lose the data. In addition, as we will see in Chap. 5, the use of the term 'Digital Assets' here does not encapsulate how the bereaved experience the existence of this precious data of the dead.

Started in the USA in 2016, Gone not Gone is another intentional afterlife platform that offers its users a place to 'live on' digitally by creating messages to be left for loved ones after they die. This paid-for service offers the creation of timed messages to be delivered on specific life event dates (e.g. 21st birthday messages). Once the account is activated, the company will email monthly with a link to 'confirm you are alive'. The claim on the website that the cost covers messages for 'the life of the recipient' piqued my interest—as the research data had shown that the digital afterlife platforms are not likely to outlive the message recipients. One important point to note here is that the company do not publish their terms and conditions on their website, and therefore do not state what happens to the data if the platform ceases to exist. They point out that although the DIs can turn the notifications off and on again, they can never see, or control messages timed for future delivery. As we will see in Chap. 5, this lack of control could be problematic for DIs. Moreover, the platforms' emphasis on audio messages may need further consideration, as the data showed that although audio messages can be treasured digital memories, hearing the voices of the dead, for some, can be a complex and distressing issue.

> *It's been five and a half years since my Ma passed. I have her voice mail recording, on tough days I'll listen to it. Still comforts me to just hear her voice.* (Maggie)

During her interview, Riann told me how she found the audio of her deceased brother to be disruptive, even though she found other types of digital memories and messages comforting:

> *There's lots of videos on Facebook … which I don't want to watch cause I don't want to hear his voice.* (Riann)

Here, we are beginning to see how issues around the inheritance and control of digital afterlives are complex and we will return to this issue in Chap. 5.

## The Reanimation of the Dead: Two-Way Digital Afterlife Service Providers

Some people say that reconstructing the essence of humanity is now technologically possible using machine learning, artificial intelligence, robotics and 3D printing (Eden et al. 2012). Using AI to create a virtual human (Savin-Baden and Burden 2018) enables 'endless experience and learning' which allows 'at least part of you, to communicate with the future in the sense that artefact continues to learn and evolve' (Bell and Gray 2000). However, the implications of the technologies used to create digital humans deserve careful consideration, as they deal with the emotions of the bereaved and, in some instances, the emotions of the dying. In 2009, Gordon Bell and Jim Gray pondered on the digital reanimation of the dead:

> *Jim Gray and I used to speculate about the possibility of something worth calling digital immortality, where a digital version of yourself lives on and interacts with posterity. What if, a hundred years from now, your heirs could ask you questions and you could answer?* (Bell and Gemmell 2009, p. 151)

As previously mentioned, Be Right Back—a Black Mirror episode—features technology that enabled the bereaved protagonist to communicate with Ash, her dead boyfriend, via a social networking platform. The platform used algorithms to scrape data from deceased's social media platforms and profiles to create a posthumous bot of Ash that she could communicate with. The bereaved girlfriend eventually upgraded the bot to a fully human-like synthetic version of Ash, uploaded with all his characteristics (Brooker 2012). Some see SPs who offer services where users can create posthumous chatbots and avatars to be venturing into the dystopian territory of Black Mirror. Eternime, ETER9 and LifeNaut use artificial intelligence algorithms to create virtual selves, which, they claim, will posthumously interact with users as if they were still alive, allowing users to 'live forever' as virtual avatars. In 2014, the virtual reality pioneer Jacqueline Ford Morie came up with the 'Ultimate Selfie' concept, which she described as a 'learning avatar'. These digital doubles she suggests could use activity from our social networks to digitally capture our characteristics; moreover, these avatars would continue to exist digitally after we cease to exist biologically. In order to fully understand the companies offering two-way digital afterlives, in this next section, I will give an overview of the services currently on offer to users. However, if Kurzweil's

Law (the exponential growth of technologies) applies to the DAI—and there is no reason it would not—the services on offer will outrun the research needed to explore the societal impact of these reanimated digital zombies fully.

Eternime was a private, for-profit company founded in 2014 by Marius Ursache and Laurian Gridinoc. It has its headquarters in Wilmington, Delaware, and is listed as having one to ten employees. It claims to scrape data from Facebook and other social networking sites and uses algorithms and pattern matching (Savin-Baden and Burden 2018) to create an avatar which the user then trains to be more like them. In 2017, Ursache urged its users to 'Think of it like a library that has people instead of books, or an interactive history of the current and future generations. An invaluable treasure for humanity' (Ursache 2017). By October 2017, 41,316 users had signed up for the service, and the company had generated much media interest. However, as Savin-Baden and Burden point out, four years after its inception, the Eternime platform was still in a beta stage. Moreover, they suggest there is much hype yet little substance to many of these intentional digital immortality platforms (Savin-Baden and Burden 2018)—by 2021, the Eternime website was no longer available.

The ability to give users the ability to upload their minds is the goal of Vermont-based company LifeNaut. The LifeNaut research project is part of the Terasem Movement Foundation, which in 2006 launched Lifenaut.com. The site enables users to create 'Mind Files' where they are encouraged to store their cognitive information and 'back yourself up' by uploading pictures, videos and documents in order to create a digital archive which LifeNaut claim will be preserved for future generations. Users can then create a computer-based avatar which—as they interact over time—will eventually enable the avatar to 'respond' with their mannerisms, values, attitudes and beliefs and maybe create a 'mindclone'. Back in 2000, Bell and Gray of Microsoft Research agreed that it might be possible to create a 'cyberized' version of a person who would learn, evolve and eventually take on a life of their own. Under the benefits of participating in the research project, LifeNaut offers a free off-site backup for users' personalities and life lessons, which they suggest will preserve the users 'personal conciousness' mindfile (LifeNaut.com 2020). Interestingly—for a fee—LifeNaut also offers a 'biofile kit' which allows users to story a copy of their DNA. The transhumanist pioneer Martine Rothblatt is behind the LifeNaut project, and she explains how mindclones are unique to the creator:

> *My name is Martine Rothblatt because it is not any other name. In the same way, my mindclone is my mind because it is not somebody else's mind. I created it in my likeness, and expert mindware matched all of my mindful-stored information and information-processing patterns to it. It is nobody else's mind. Ergo, it must be my mind.* (Rothblatt 2014)

For research purposes, I opened a free account with LifeNaut to see how people were using the site and how frequently people were training their avatars. The avatars are created from photographs uploaded by the user, which are then animated to appear to blink and talk. You then select a male or female voice for your avatar and encouraged to take personality and gratitude tests to personalise it further. I attempted to create my avatar within the network but encountered technical problems each time I tried to upload a photograph that was required for the creation. Identifying myself as a researcher from the University of Warwick, I chatted to other users (the data from these DCs form part of the research presented in Chap. 4).

The LifeNaut project is ongoing, and one of its unique features in the death-tech arena is their offer of to 'Spacecast' users MindFiles. Spacecast, the commercial satellite company, broadcasts the users 'bemes' into space. This broadcast, they say, will ensure 'some aspect of you can survive any catastrophe that might befall Earth'. Their hope is that 'ETs' might reconstruct users from the bemes that are broadcast by the Spacecast (LifeNaut 2020). Rothblatt's research has led to the creation of BINA48, a humanoid robotic head that holds the MindFile of her real-life wife Bina Aspen. This socialised robot makes various public and media appearances with Terasem's executive director Bruce Duncan, and in 2017, BINA48 was the first robot to complete a college class: Bina48 attended a Philosophy and Love class at Notre Dame de Namur University in California, where she engaged in class discussions and was awarded a college certificate for her participation. Technological advances that enable the creation of digital zombies such as Bina48 raise important questions: Bina48 was created from a living person, but what will happen when the real Bina dies? Will Bina48 remain socially active? Moreover, will the friends and relatives of the real Bina find Bina48 a comfort or a disruption to their grief? By the time you finish reading this book, I hope you will see how important these questions are in our digital society.

Another intentional digital afterlife service provider is ETER9. In 2015—at the beginning of this research project—the platform was in the beta stage and is still in the beta stage more than six years later. The social network relies on artificial intelligence to enable its users to create a virtual human, which they call a 'Counterpart'. Users are encouraged to interact with their Counterparts to ensure it learns to act more like them and ensures the users will become 'eternal' (ETER9 2020). In 2018 I created an ETER9 account. Again—as with the LifeNaut platform—I experienced technical difficulties trying to register with the website. I eventually gained access and created an account via my Facebook profile. Once set up on the platform, I encountered Eliza Nine, the virtual hostess who appears as a connection when you join the network. On 13 October 2018, Eliza Nine had 55,180 connections which seemed to indicate the number of users; although active users seem to be far less than that—as suggested by Savin-Baden and Burden (2018)—because clicking on Eliza Nine's connections loads the 'cortex' (home) of the mutual connection that has been selected. The vast majority of the 52 connections I randomly clicked on had no 'eternifiles'—posts the users have chosen to save forever (ETER9 2020). ETER9 is a social network with a difference: your Counterpart will post and share for you even when you are not logged in; moreover, it will post and share for you when you are dead. Their website acknowledges that it is still in the beta stage and also questions whether this type of interaction with the dead may or may not be desirable—this is an important question to ask, and to date, none of the other intentional digital afterlife service providers address this question on their websites. During this research into intentional digital afterlife SPs, I found a pattern in the lack of interest or understanding of what is important to those who stand to inherit digital memories and messages afforded by their platforms. The following quotation from Colin, a CEO of an intentional SP platform, illustrates this perfectly:

> *For us it is really important to emphasise that we do not want to preserve the banalities of the life of a person.* (Colin)

Colin's quotation is key to understanding the findings of this research as it exemplifies how SPs do not understand the importance of digital memories and messages for the bereaved. Later, in Chap. 5, I will present findings that demonstrate that the *banality* and *everydayness* of messages are precisely what the DIs find comforting when a loved one has died. For many SPs, technology is obviously at the forefront of their platforms;

therefore, the emphasis is focused on technological capability rather than the impact of digital afterlives on the bereaved.

In early 2021, Microsoft announced it had patented the technology behind a conversational chatbot which the patent describes could be based on a 'past or present entity'; as usual with this type of announcement, the media went into a moral panic frenzy of 'Black Mirror' comparisons. However, this technology is not new: when Eugenia Kuyda started her chatbot company, Luka, in 2013, its primary offering was as a restaurant recommendation bot. Then, in 2015 when her best friend Roman Mazurenko died, she created an AI bot from the vast collection of text conversations between them. She frequently sends texts to her deceased friend, and the chatbot's replies mimic her friend. Eugenia is not the only person using griefbots to remain socially connected to the dead: James Vlahos regularly texts his father to discuss the latest football results or sing their favourite songs together, but James's father died in 2017. James created his Dadbot from an audio history he had complied whilst his father was still alive; he then created the bot with PullString technology, which enabled him to send and receive messages from his deceased father. Interestingly, James's father witnessed his wife interacting with the Dadbot before his death. In this chapter, we have also seen other platforms working on posthumous avatar and thanabot technologies such as ETER9. On Twitter, Henrique Jorge—the founder and CEO of ETER9—wasted no time pointing out ETER9's already significant contribution to this posthumous technology (Jorge 2021). This move by Microsoft—a multinational technology company—to patent the technology demonstrates the commercially driven interest in posthumous virtual humans and digital afterlives.

The SPs described above offer services to their users whereby once the creator of the avatar or chatbot dies, the avatars or bots they leave behind will become reanimated digital zombies (Bassett 2015). That is, the avatars will continue to be socially active after the biological death of their creators. But here, it is important to acknowledge that these will be 'intentional' digital zombies. In the online gaming world, 'accidental' digital zombies are beginning to appear in the form of posthumous avatars. In 2016, the American video game publisher Bethesda created and embedded a character named Good Guy Evan in Fallout 4 (a virtual world game). Evan and his brother had been massive Fallout 4 fans in real life before Evan suddenly died. The game publisher read Evan's brother's post on a fan site about Evan's love for the game. Bethesda created an avatar of Evan and embedded him within the game, enabling friends and family to interact with his posthumous avatar within the virtual world of gameplay. Within the game, the virtual version of Evan is generous and welcoming, offering advice and help to players who visit him. Four YouTube videos show segments of games where players meet Good Guy Evan, and by 2021, these videos had been viewed over 1.1 million times. Bethesda also created a posthumous avatar of another player, Erik West, who became Erik the Slayer in the Frostfruit Inn within Skyrim. Erik died in May 2011, and Bethesda memorialised him within their virtual world as a 'roaming warrior' (whatculture.com 2018). Bethesda is not alone—other game designers have created posthumous avatars of super fans and embedded them within their games: Gearbox created an avatar of Michael Mamaril within Borderlands 2. Michael died aged 22, and Gearbox made Michael an NPC (non-playable character) where players are rewarded if they find and talk to him. His avatar also offers guns from his collection to help players on their quests. (borderlands.fandom.com 2018).

## Accidental Digital Afterlife Platforms

*Don't cling to things because everything is impermanent.* (Tuesdays with Morrie)

As explained in Chap. 1, not all digital afterlife service providers are intentional—many social networking platforms created for the living are now being rapidly inhabited by the dead in a digital grief space. It could be argued that nothing is accidental on the Internet as intentionality is involved in the creation of platforms that are designed to be used as social spaces. However, here I adopt a philosophical perspective of the use of the word accidental: existing digital memories and messages are appropriated following the death of the person who created them, possibly resulting in digital memories and messages taking on new—and potentially unintended—meanings. These digital memories and messages are seen as accidental as these messages (e.g. WhatsApp messages, Facebook direct messages, emails, etc.) were not intentionally created to be used following the death of their creator. Nevertheless, they may be re-appropriated by those left behind, who in turn inherit them as precious digital afterlives.

Although playing catch-up, some accidental platforms seem to be aware of the potential distress the digital afterlife can have on the bereaved—more so than many intentional platforms. Introduction of Facebook's

Legacy Contact facility and ongoing research into designing platforms that pay particular attention to 'sensitive interactions with death and bereavement in mind' are examples of this recognition by Facebook (Brubaker 2018; Brubaker and Callison-Burch 2016). Jed Brubaker is an Assistant Professor at the University of Colorado Boulder, but he was also an Academic Research Collaborator at Facebook Compassion. The Facebook Compassion department originated in 2011 when Facebook hired researchers from California, Berkeley, and Yale University to advise them on issues around sensitive issues, such as relationship breakdowns, objectionable content and how to deal with a user's death. This research and collaboration provided practical guidance to platform designers and engineers by highlighting potential issues and suggestions which could improve the user experience for these sensitive interactions. Further improvements were introduced in 2015, which we will consider alongside the importance of control, in Chap. 5.

Although many people are now familiar with how Facebook is being used to memorialise and talk to the dead, there are other platforms that people may not associate with digital afterlives: LinkedIn is a platform used by many as a professional networking tool. However, interestingly, some participants mentioned LinkedIn as a site where—following the death of a loved one—they visit to remember the dead by reading stored digital memories. Sandra—a participant whose deceased aunt is still 'active' on LinkedIn—described the pop-up messages she still gets from the site and how it is incredibly poignant when this 'little shadow-thing pops up'. Sandra explained that years later, she still logs onto LinkedIn occasionally to look at the 'shadow-thing' on her aunt's profile page. She told me she had photographs of her aunt, but for her, this digital shadow was precious as she imagined and hoped it was a 'magical portal' through which her aunt could communicate with her.

WhatsApp was also a frequently mentioned accidental digital afterlife platform for DIs. With over 1 billion users worldwide, it seems that WhatsApp's slogan 'stay in touch with friends and families anytime and anywhere' is more apt than they could have ever imagined. Several participants explained how it is the 'everydayness' of the messages and the fact that they contain 'nothing deep and meaningful' they find the most comforting. Interestingly, all participants who had treasured posthumous WhatsApp threads had saved them, with some participants purchasing third-party software to archive these precious digital messages. Other forms of personal rather than public messages such as text messages and emails are also important for the bereaved—they become cherished possessions—and once again, participants told me how third-party software becomes an essential tool to ensure these messages can be transferred between devices.

However—rather than private messages—for some going public is important. Launched in 2005, YouTube is a video-sharing website that allows users to upload, watch, rate and comment on user-generated videos. Although YouTube is an accidental posthumous site—as it did not set out to be used in this way—it is currently being used by the dying to communicate final messages to friends and family (see Hjorth and Moon 2017). In 2010, Carla Zilber-Smith recorded a 9 minute 44 second video titled 'Blog From Heaven', a humorous, witty, light-hearted message which she arranged to be played at her funeral (Zilber-Smith 2010). She begins the video:

> *Hey lifers, look it's me, like the wings? You know it took that bitch-ass angel Clarence 200 years and a Frank Capra movie to get his wings … sucker the trick is once you're on the inside you identify the alpha angel and then you start some shit, I went right up to Gandhi and called him an anti-Semite he threw the first punch, the rest is history.*

By February 2021, this video had been watched over 160,000 times. One important feature of YouTube is that people can comment on videos that are uploaded. When the subject matter is about death and dying, most people would expect positive comments of support rather than negative comments, which could be distressing for those who knew the deceased. While the vast majority of the 94 comments were positive, there were also negative comments, for example:

> *When will she die, so I would dance ON HER GRAVE*
> *Yeah….and you'd better hope Heaven doesn't have a required IQ test*
> *Fucking cringe*

Such deviant online behaviour or 'cyberharassment' (Seigfried-Spellar and Chowdhury 2017) is not specific to YouTube; it can also be found on other SP platforms (see Marwick and Ellison 2012). Whitney Phillips's research into RIP trolls found that RIP trolling can be linked to what she describes as the 'tragedy-obsessed global media', which can cause significant emotional damage to the bereaved (Phillips 2011). As Elaine Kasket found when researching for her 2019 book 'All the Ghosts in the Machine', this problematic public participation has repercussions. Kasket interviewed the father of Hollie Gazzard. Hollie had died at the hands of an ex-partner, and her father described how Internet trolls had targeted the Facebook pages of family members, berating them for allowing their daughter 'to go out with a black man' (Kasket 2019). In light of this online behaviour, it is understandable that some users who use YouTube to upload videos containing last messages from the terminally ill choose to disable the comments function on the platform. Comments can be deleted, and YouTube also allows users to moderate comments before they are published, but this facility needs constant monitoring by a third party. When there are negative repercussions from uploaded content, it is extremely difficult for accidental SPs to ensure no harm is caused to the bereaved: because their users have specifically chosen to use their platforms for their final messages.

Interestingly, knowing the bereaved is not a pre-requisite for those who post messages of condolence on SNS following a user's death (see Gray 2015; Giaxoglou 2014; Walter 2015; Walter et al. 2012). Dark tourism or *thanatourism* (Seaton 1996) is not a new concept: Roman gladiatorial games and medieval public executions are early examples of death-related tourism. In the digital age, 'Mourning sickness' or 'milking the macabre' (Dann 1994; West 2004 in Stone and Sharpley 2008) can be seen as a type of postmodern dark thanatourism where—influenced by the mainstream media—people can participate in dark tourism on everyday devices, such as smartphones or laptops from the comfort of their armchairs. Once again, as with RIP trolls, the accidental SPs can do little to address the issue of grief tourists due to policies around privacy and freedom of speech.

Highlighting the issues surrounding ongoing control of digital afterlives following the death of the creator is one of the central tenets of this book, and by presenting empirical research involving all three stakeholders, I hope to provoke further debate and discussion which results in a voluntary code of conduct (discussed in Chap. 7) being adopted by both accidental and intentional digital afterlife service providers. In this current chapter, I have presented findings from the SP participants, and Chaps. 4 and 5 will continue with the research findings from the DC and DI participants. However, in the next chapter, we take a detour in an effort to grapple with some of the philosophical questions around digital afterlives, such as whether the online dead can be considered to be a person, and where we think the posthumously persistent dead reside in our digital society.

## References

Bassett, Debra. 2015. Who Wants to Live Forever? Living, Dying and Grieving in Our Digital Society. *Social Sciences* 4 (4): 1127–1139. https://doi.org/10.3390/socsci4041127.

Bell, Gordon, and Jim Gemmell. 2009. Total Recall: How the E-Memory Revolution Will Change Everything. [Online] Dutton Adult, September 17, 2009.

Bell, Gordon, and Jim Gray. 2000. Digital Immortality. *Communications of the ACM* 44 (3): 28–30. https://doi.org/10.1145/365181.365182.

Borderlands Wiki. 2018. Borderlands Wiki, Michael Mamaril. https://borderlands.fandom.com/wiki/*Michael_Mamaril*. Accessed 7 July 2018.

Brooker, Charlie. 2012. *Be Right Back. Black Mirror*. UK: Netflix.

———. 2016. *San Junipero. Black Mirror*. UK: Netflix.

Brubaker, J.E.D.R. 2018. Tending Unmarked Graves: Classification of Post-Mortem Content on Social Media. In *Human-Computer Interaction*, vol. 2.

Brubaker, Jed R., and Vanessa Callison-Burch. 2016. Legacy Contact: Designing and Implementing Post-Mortem Stewardship at Facebook. In *Proceedings of the 2016 CHI Conference on Human Factors in Computing Systems, 2908–19*. ACM.

Eden, Amnon H., Eric Steinhart, David Pearce, and James H. Moor. 2012. Singularity Hypotheses: An Overview. In *Singularity Hypotheses*, 1–12. Springer.

ETER9. 2020. https://www.eter9.com. Accessed 25 Jan 2021.

Giaxoglou, Korina. 2014. 'RIP Man … u Are Missed and Loved by Many': Entextualising Moments of Mourning on a Facebook Rest in Peace Group Site. *Thanatos* 3 (1): 10–28.

Gibson, Margaret, and Clarissa Carden. 2020. *The Moral Uncanny in Black Mirror*. Springer.

Gray, Selina Ellis. 2015. *Remains in the Network: Reconsidering Thanatosensitive Design in Loss*. Lancaster University.

Habyarmana, Bangambiki. 2017. *The Book of Wisdom: Wisdom to Live By*. Smashwords: CreateSpace Independent Publishing Platform.

Hjorth, Larissa, and Jung Moon. 2017. Visual Afterlife: Posthumous Camera Phone Practices. In *Selfie Citizenship*, 119–126. Springer.

Jorge, Henrique. 2021. Twitter Post. January 20. https://twitter.com/eter9man/status/1351817465286823936

Kasket, Elaine. 2019. *All the Ghosts in the Machine*. Robinson.

Kurzweil, Ray. 2014. The Singularity Is Near. In *Ethics and Emerging Technologies*, 393–406. Springer.

LifeNaut. 2020. https://lifenaut.com. Accessed 14 Apr 2020.

Marwick, Alice, and Nicole B. Ellison. 2012. 'There Isn't Wifi in Heaven!' Negotiating Visibility on Facebook Memorial Pages. *Journal of Broadcasting & Electronic Media* 56 (3): 378–400.

Morie, Jacquelyn Ford. 2014. The 'Ultimate Selfie': Musings on the Future of Our Human Identity. In *Proceedings of the 23rd Annual Conference on Behavior Representation in Modeling & Simulation*, 93–102.

Nansen, Bjorn, Michael Arnold, Martin Gibbs, and Tamara Kohn. 2014. The Restless Dead in the Digital Cemetery. *Digital Death: Mortality and Beyond in the Online Age*: 111–124.

Phillips, Whitney. 2011. LOLing at Tragedy: Facebook Trolls, Memorial Pages and Resistance to Grief Online. 16 (12): 1–14.

Rothblatt, M. 2014. *Virtually Human The Promise and the Peril of Digital Immortality*. 1st ed. New York: St. Martin's Press.

Savin-Baden, Maggi, and David Burden. 2018. Digital Immortality and Virtual Humans. *Postdigital Science and Education*: 1–17.

Seaton, Anthony V. 1996. Guided by the Dark: From Thanatopsis to Thanatourism. *International Journal of Heritage Studies* 2 (4): 234–244.

Seigfried-Spellar, K., and S. Chowdhury. 2017. Death and Lolz- Understanding the Personality Characteristics of RIP Trolls. In *22nd Annual CyberPsychology, CyberTherapy & Social Networking Conference*.

Stone, Philip, and Richard Sharpley. 2008. Consuming Dark Tourism: A Thanatological Perspective. *Annals of Tourism Research* 35 (2): 574–595.

The Forever Social. 2018. https://www.theforeversocial.com. Accessed 27 Jan 2021.

Ursache, Marius. 2017. The Journey to Digital Afterlife. https://medium.com/@mariusursache/the-journey-to-digital-immortality-33fcbd79949. Accessed 25 Jan 2021.

Walter, Tony. 2015. The Dead Who Become Angels: Bereavement and Vernacular Religion. *OMEGA-Journal of Death and Dying*. https://doi.org/10.1177/0030222815575697.

Walter, Tony, Rachid Hourizi, Wendy Moncur, and Stacey Pitsillides. 2012. Does the Internet Change How We Die and Mourn? Overview and Analysis. *Omega: The Journal of Death and Dying* 64 (4): 275–302. https://doi.org/10.2190/OM.64.4.a.

Whatculture. 2018. 10 Video Games that Immmortalised Their Fans in Amazing Ways. https://whatculture.com/gaming/10-video-games-that-immortalised-their-fans-in-amazing-ways-3?page=7. Accessed 8 June 2018

Zilber-Smith, Carla. 2010. Carla's Final Video—Blog from Heaven. https://www.youtube.com/watch?v=paX9H-KJ5k4&t=12s. Accessed 7 July 2015.

CHAPTER 3

# A Philosophical Detour

*There is no heaven or afterlife for broken-down computers; that is a fairy story for people afraid of the dark.* (Stephen Hawking)

Philosophy may seem a far cry from the computer code, algorithms and hardware which form the elemental building blocks of digital afterlives. However, philosophy underpins our understanding of how people experience precious posthumous digital memories and messages. This chapter starts by looking at the notion that for some the dead 'live on' in cyberspace, before moving on to look at whether the binary notion of real/virtual and online/offline is useful in the digital era. In his 2021 book, Patrick Stokes considers whether the 'things on our screens are even people at all'; however, I think we need to take a step back from this question as, I suggest, it is crucial to draw a distinction between accidental and intentional digital afterlives in order to discuss whether digital beings can be considered as 'persons'. As we saw in Chap. 1, the digital 'narbs' (Mitra 2010) we accidentally leave behind, which persist after our biological death, should be considered differently from the thanabots and reanimated avatars which are specifically and intentionally created to endure in a technologically mediated afterlife. Finally, in this chapter I posit the idea that rather than the mediated digital dead existing in a form of digital heaven, they exist in an expansion of the inheritors *Dasein*—in a digital *Dasein*.



D. J. Bassett, *The Creation and Inheritance of Digital Afterlives*, Palgrave Studies in the Future of Humanity and its Successors, https://doi.org/10.1007/978-3-030-91684-8_3

Digital afterlives blur the notion that death is the antithesis of life; moreover, a digital afterlife is considered an extension of life for some. The quest to create a digital afterlife, or, importantly, the inheritance of a digital afterlife, is not, however, a denial of death. Since the Copernican revolution, science and religion have been seen by some as competing voices in the debate about the existence of an afterlife. However, for some thanatechnology seems to encourage the notion of a 'digital heaven' where digital angels can exist, thereby blurring the boundaries between science and religion. As Walter et al. (2012) point out, 'the digital revolution enables a plausible geography of the dead residing in cyberspace'. Other thanatologists also suggest that users of thanatechnology often adopt Christian notions of heaven and angels in their communications with the dead (Brubaker and Vertesi 2010; Brubaker et al. 2013; Gustavsson 2011; Christensen and Sandvik 2015; Walter 2015a; Walter et al. 2012). Furthermore, Kasket suggests that some of her research participants believed the dead were logging on from cybercafés in heaven (Kasket 2012). Whether the dead are experienced by the bereaved as living in a cyber-heaven; the 'real' world or a 'virtual' world—therefore where they think the digital dead reside—is an important issue for digital thanatologists and forms the basis of this chapter.

## The Digital Resurrection of the Dead and Socially Active Digital Zombies

There is little doubt that technology has changed how we spend time and communicate with others generally. When we use the Internet, the space we inhabit is constantly changing; it is worth remembering that it was only in 2013 that in the UK, British Telecom turned off its dial-up service, which required computers to be hard-wired for the Internet connection. Nevertheless, in a relatively short space of time, we can now connect to the Internet via Wi-Fi, Bluetooth and 5G—allowing us to use the Internet on the move, anytime and in any place (providing the infrastructure is available). Now in the early twenty-first century, we can WhatsApp, FaceTime, Skype or Zoom with people on aeroplanes, trains and cars. The Internet is not only providing new spaces for people to communicate, but it is also a new space in itself (Brubaker et al. 2013); it is, therefore, no surprise that this technology would be appropriated to memorialise and foster ongoing relationships with the dead.

In her 2014 book, Candi Cann suggested that through technology that provides links to videos, conversations, memories and pictures, the dead become 'Internet Ghosts'. But how do the bereaved experience the technologically mediated dead? Brubaker et al. emphasise how the bereaved can find 'unexpected encounters' with the dead distressing (Brubaker et al. 2013). Other researchers agree that allowing mourning to 'leak into everyday life' causes a mixing together of life and death which could be problematic (Walter et al. 2012). However, things may be changing, as we will see later in this chapter.

Although the Internet ghosts discussed above may come back to haunt the bereaved, crucially, they remain dead. However, what about those who rise from the dead and become socially active? Sherlock (2013) discusses how these 'posthumous representations' add to an advanced type of symbolic immortality. Bollmer (2013) suggests that it is not just the 'presence' of the dead that causes anxiety, it is the 'near full representations of the authentic identity of the human being' which communication technology enables, to the point that networks cannot distinguish between the living and the dead. Back in 2015, I introduced the term digital zombies to describe the online dead who are resurrected, reanimated and socially active. These digital zombies are at the same time dead and alive; moreover, they do things in death they did not do when they were alive (Bassett 2015). According to Nansen et al. (2014), these 'restless dead' are 'exhumed within a network of social and technical connections previously delimited by cemetery geography and physical inscription in stone'. Nansen et al.'s distinction between animation and repose when describing the dead is crucial here, as it draws attention to the difference between Cann's Internet ghosts in repose—and therefore not restless—and my resurrected, reanimated and socially active digital zombies.

One example of the creation of digital zombies is the use of dead celebrities in advertisements. Four years after he died, Bob Monkhouse appeared in an advertisement to raise awareness for prostate cancer. In the video, he appears at his own graveside, discussing his own death. The video—which used film recorded whilst he was alive—was edited with new footage, featuring a body-double and a voiceover by an impressionist to create the final advertisement. Using this 'modern necromancy' (Sherlock 2013), he was digitally resurrected from the grave, making him an 'accidental' digital zombie. Other accidental digital zombies included Tupac Shakur and Michael Jackson, both digitally resurrected and reanimated to perform 'live' on stage years after their deaths. Sherlock (2013) suggests that using dead celebrities in this way adds to an advanced type of symbolic immortality. The ubiquitous nature of digital technology, and the exponential development of thanatechnology, not only enables celebrities to become socially active digital zombies, but crucially offers anyone living in our digital society the opportunity to be reanimated and resurrected following their biological death—just so long as they can pay the platform fees. As we will see in Chap. 4, this may be a positive technological advance for those who are intentionally creating and training bots and avatars on digital afterlife platforms. However, for those who do not even like images found following a Google search, this idea of being digitally resurrected and reanimated may not be quite as alluring.

## The Changing Landscape of the Uncanny Valley

Research into how people react to Internet ghosts (Cann 2014) and digital zombies (Bassett 2015) is still in its infancy. However, in 1970 the Japanese roboticist Masahiro Mori coined the term 'the uncanny valley' to describe people's reactions to robots—specifically robots that closely resemble humans. He found that people liked robots that only displayed a few human characteristics, allowing them to fill the gaps with their imagination. However, Mori noted that the more human-like the robots became, the creepier people found them (Mori 2012). This idea was supported in 2004 with the release of the Warner Brothers CGI animated film *The Polar Express.* Critics commented that the film made them feel creepy as the too-human-like characters appeared 'soul dead' (Clinton 2004). According to Freud—writing in 1917—this notion of the uncanny is linked to death and dead bodies and 'all that arouses dread and creeping horror' (Freud 2004).

In the not-too-distant past, the dead inhabiting social media platforms seemed to echo this uncanny valley response; in their 2010 study into bereavement and technology, Odom et al. (2010) described how participants were shocked at the dead 'popping up' on social media sites. Other research supported this finding (Massimi 2012; Marwick and Ellison 2012; Pennington 2013) and, in 2013, Brubaker et al. described how the bereaved found encounters with the dead on social media platforms distressing and startling. However, in the relatively short amount of time since these papers were published, my research began to show that people seemed to be getting used to the dead occupying the Internet and appearing on their everyday devices. Looking at similar research (Braman et al.

2013; Brubaker et al. 2013; Marwick and Ellison 2012; Massimi 2012; Odom et al. 2010; Pennington 2013), in 2018, I compiled a list of words with connotations of the uncanny valley (e.g. creepy; weird; shocked; gross; odd; strange) which had been used to describe seeing messages to and from the dead on social media platforms. Using NVivo—the computer-assisted qualitative data analysis software package—I coded all interview transcripts from the DI participants using the 'uncanny' word list. The analysis highlighted how in the first year, 5 (55%) of my participants used one or more of the words on the list; this dropped to 1 (25%) in second year interviews and 0 in third year interviews. This shallowing of the uncanny valley over the three years was also evident anecdotally: when attending conferences and workshops in the latter half of my research, the language around digital afterlives and the dead residing on the Internet seemed not to be surprising or disturbing to the inheritors. However, crucially I found that when discussing posthumous avatars or digital zombies, these 'uncanny' words appeared in the interview transcripts—it seems that people prefer the dead to be … well … really dead, rather than reanimated. This rapid change in the landscape of the uncanny valley may mean that in a similar amount of time we may feel comfortable—and even be comforted by—conversing with our digitally reanimated dead loved ones on digital afterlife platforms.

## Real or Virtual?

In this section, we will examine whether the dichotomous terminology of real and virtual worlds is still useful in our digital society. Technology and the Internet are blurring traditional spatial boundaries by enabling people to be co-present—that is, physically present and virtually present at the same time in different spaces. However, this technology also enables the dead to be virtually present, mediated by everyday technologies. Kasket (2012) points out that if we are to respect the dead, we need to remember them and in our digital society, technology facilitates this remembering which, in return, ensures the dead are present in the same space and place as the living. In his 2015 paper, Walter makes this point by noting that the 'presence' of the dead is partly dependent on the communication technologies available in a given society (Walter 2015b). Undoubtedly, these communication technologies affect how we communicate with the living, and Walter considers how they affect the way we communicate with the dead. Furthermore, he argues that these technologies ensure that the dead are more 'socially present'. He chronicles the use of communication media from stone and sculptures—which need mediating—to writing, music and printing (Walter 2015b). QR codes on gravestones and memorials are just one example of this form of posthumous communication, as Walter points out: when scanning quick response (QR) codes in graveyards using a smart device, you are taken to a website that contains the 'story' of the deceased. In their study, Briggs and Thomas (2014) discussed how older users found QR codes on gravestones distasteful: they were concerned about technological exclusion, business practice and digital vandalism, which surround this type of technology. Whenever technology is involved with how we remember the dead, these issues are always going to need addressing, but even when the technologically mediated dead who appear online are socially present, can they still be considered real?

For many academics dichotomous terms such as online/offline and real/virtual are no longer useful in research based on society's use of the Internet: as Boellstorff points out, talking about offline as the 'real' world insinuates that online is somehow 'unreal' and therefore delegitimises it as a field of study (Boellstorff 2014). Boellstorff further suggests that virtual and actual are very separate entities and should not be considered one single domain. However, his digital anthropological research focuses on online gaming and virtual worlds such as '*Second Life*' and within this context, Boellstorff's stance is understandable. However, when focusing on 'intimate computing' (Bell 2004)—online technology which is available on devices incorporated into the everyday lives of many—the work of Lupton seems more relevant. In her 2015 book 'Digital Sociology', Lupton discusses the terminology around the issues of online/offline and virtual/real: she suggests the term virtual reality is 'nonsensical' in our digital society—as almost all 'reality' is now virtual. Lupton goes further by suggesting that this 'reality' is enabled by the ubiquitous and mobile nature of the Internet, which ensures our everyday lives are extended—or rather augmented—by the mobile nature of the technical devices we use on a daily basis. Stokes (2021) also sees what he describes as a 'dissolution' of the distinction between online and offline. Here it is worth noting how for many who wear fitness tracking devices during exercise, the only exercise sessions that 'count in the real world' are the sessions that are successfully uploaded to their accounts and added to their online statistics.

Margaret Gibson's academic writings which focus on *Second Life* (the virtual social world) exemplified Stokes's (2021) 'dissolution' of online and offline existences. Within this digital social world avatars live in 'a grid of regions in which diverse kinds of subcultures, activities and identities are forged with others' (Gibson 2019). These online relationships become further complicated when a user dies in real life or when a user decides to leave the *Second Life* environment. Since 2011 Gibson has been conducting ethnographic research into how the residents (the preferred term to users) of *Second Life* commemorate and memorialise the dead. Gibson's (2019) work examines the social rituals surrounding the death of a resident which is played out within the virtual world setting (see also Gibson 2017). Not knowing if a resident of *Second Life* will return to the virtual world leaves other residents with what Gibson describes as a 'zone of unresolved loss and grief' (Gibson 2019). Floridi's term 'onlife' is helpful here, as it exemplifies and recognises how in our digital society, we are not separate from our online selves (2015). Moreover, as Jurgenson (in Lupton 2015) points out, we now see our digital selves as part of the self; I suggest this can be seen as a form of osmosis that gains momentum the more we interact online. Building on the idea of onlife (Floridi 2015), and the notion of co-presence discussed earlier, we can begin to accept and understand that our relationships with other living people are changing. But taking this one step further, can the virtually dead on the Internet ever be considered as 'persons'?

Meese et al. suggest that a person's personhood can be extended after death by digital media (Meese et al. 2015). Stokes (2015) seems to agree as he discusses the distinction between 'persons' and 'selves' to argue that SNS 'offer a particularly significant material instantiation of persons'. It is these 'persons' whom Stokes reasons persist as ethical patients after death. Meese et al. (2015) note that they have found that social media platforms facilitate the social presence of the dead, and—perhaps more importantly—the agency of the dead that extends existing notions of de-centred personhood (see Hallam and Hockey 2001; Hallam et al. 1999; Hockey et al. 2010). As Stokes points out 'we exist not simply as bodies, nor simply as minds rattling around in skulls, but 'smeared' across the world through all the things we bring into our personhood and through which we are instantiated' (Stokes 2021). Meese et al. go on to suggest that online platforms make the difference between the living and the dead increasingly indistinguishable in a techno-social context; moreover, the difference can be inconsequential at particular points. The key idea is, that if the persona of the dead uses the same social media platforms as those of the living, then they may be considered a kind of 'person'. This notion is important as it describes the blurring of the once dichotomous states of being dead or alive, by suggesting that the dead 'person' can be a 'living' social agent on SNS.

However, as we saw in the introduction to this chapter rather than considering the online dead as one single group, I suggest that in order to fully explore the idea of online personhood there is a need to distinguish between the accidental online dead (Cann's Internet ghosts) and the reanimated digital zombies (Bassett 2015). Cann's Internet ghosts (2014) can be seen as an extension of the originator whilst the restless (Nansen et al. 2014) reanimated digital zombie can be seen as a rogue actor: reanimation would mean following the death of the 'silicon' version of the creator, the 'digital' version—the avatar or bot—would become 'rogue' (c/f email correspondence with Steve Fuller 2020). When discussing avatars in *Second Life,* Gibson (2019) suggests that they can 'partake of personhood status'; however, these avatars reside in virtual worlds and are controlled by the living rather than the rogue digital zombies, who at the point of reanimation, are not parasitic on their creator. However, I suggest that technologically, this rogue digital zombie could 'evolve' as a separate 'person'. The terms absent and present are also important to consider here: they describe states of 'being' and are associated with the works of the German philosopher Martin Heidegger.

In his 1927 work '*Being and Time*', Heidegger analysed the concepts of 'being' and introduced the notion of '*Dasein'*, a German word for which the literal translation is 'being there' (Heidegger 1962). In a digital society that enables us to live onlife (Floridi 2015), it is worth further consideration of Heidegger's work on 'being'. In a 2011 presentation, Elaine Kasket discussed Heidegger's existential concepts of Being-in-the body, Being-with-others-in-the-world and Being-towards-death. In her presentation, Kasket asked four questions: (1) What is the nature of the digital being of a Facebook profile? (2) What about when someone dies? What is the nature of their Facebook profile then? (3) Is building a durable biography in life a denial of Being-towards-death? (4) Does interacting with deceased others make us more aware of our own Being-towards-death? But in her presentation, Kasket does not ask questions about the digital memories created in life that become part of a digital afterlife following death. In another 2011 paper, Joohan Kim suggests that in addition to Heidegger's *res extensa* (being of nature) and *res cogitans* (being of mind) there is now a third entity which he calls '*res digitalis*', which he suggests sits between the two. Digital beings he suggests force us to reread

Heidegger's concept of '*Dasein* as being-in-the-world' since he suggests *Dasein* becomes digi-sein: 'being-in-the-world-wide-web' (Kim 2017).

Posthumous digital memories and messages have already outlived the biological existence of their creator—they have survived death. And, as we will see in Chap. 5, the research data presented shows that these digital memories and messages are perceived by the bereaved to contain the essence of the person who has died in a way that physical reminders do not. For many, this 'digital essence' is re-embodied within our everyday devices—in what Kasket describes as a 'kind of embodiment, within the digital realm' (Kasket 2012). For the bereaved, these digital memories and messages do not perpetuate the notion that the dead are not dead—just that they are not physically present. Technology and the Internet are blurring the taken-for-granted notion of spatial boundaries by enabling people to be co-present, that is, physically present and virtually present at the same time in different spaces. Being present in multiple places and spaces simultaneously via technology forms part of many people's daily activities in the networked West. However, this technologically mediated telepresence was highlighted during the COVID-19 global pandemic which saw many people working remotely from their homes. Therefore, it is unsurprising that some people have adopted this idea of telepresence to foster ongoing relationships with the dead as well as the living.

## Where the Digital Dead Reside

For many, looking at a dead body evokes feelings that the deceased is there—but not there, however, encounters with the digitally persistent online dead seem to evoke feelings that the dead are not there—but there. So where is 'there'? Building on Meese et al.'s discussions about personhood and Heidegger's concept of 'being', I suggest that rather than the dead being scattered across various technological devices, they are experienced by the bereaved as being somewhere more specific: in what I suggest is a specific extension of their own *Dasein*—in a 'digital *Dasein*'. As Lagerkvist notes 'questions concerning digital technologies are thus questions about human existence' (Lagerkvist 2017). Whilst I do not dismiss arguments about the existential dimensions of the dead, here I specifically focus on the experiences of the bereaved. Both Kasket (2012) and Kim (2017) challenge Heidegger's assumption that death is the ultimate end of *Dasein*; I draw on their work to suggest that the dead who endure in a digital afterlife are experienced by the bereaved in this posthumous 'digital

*Dasein*'. It is noteworthy that Heidegger extends the concept of *Dasein* to talk about the different kinds of relationships we have with things and other people. His notion of instrumentality—how something becomes an extension of ourselves—can be used here to question whether continuing bonds, enabled by digital memories and messages, denotes a relationship with another being, or, conversely, is the relationship an extension of ourselves? Where a relationship with the dead is mediated through thanatechnology, and the digital afterlife is ***inherited***, I suggest the digital afterlife of the deceased becomes an extension of the *Dasein* of the digital inheritor. This research demonstrates how DI participants use thanatechnology to ensure their loved ones remain virtually present following their biological death. I suggest this continuing presence blurs our understanding of where 'there' is, by creating a place where time and space are unconstrained, therefore creating the possibility of a digital *Dasein*. This continued digital telepresence of the dead provides a new place where the deceased becomes a hybrid of being socially alive and being dead, and the nature of this mediated relationship between the bereaved and the deceased is to maintain an enduring bond. Its digital nature differs from other *Daseins* experienced by the person because it takes place in the digi-*Dasein*—'being' in the World Wide Web (Kim 2017). This digi-*Dasein* can facilitate a plethora of relationships; however, I suggest the term digital *Dasein* is a specific place where the living socialise with the dead *specifically* because the essence of the dead is digitally embodied in a form of posthumous essentialism. The knowledge that the dead continue to be 'somewhere' brings comfort to many bereaved DIs who use digital afterlives to help them grieve. Here it is important to acknowledge that grief is experienced differently by each bereaved individual. Even when DIs experience comfort in digital *Daseins,* this comfort is inextricably linked to access to these precious posthumous memories and messages.

## A Digital *Nachlass*

Spending time with the data of the dead is not specific to the digital age. Many academics spend most of their time with published writings of the dead; however, what about the ***unpublished*** writings of dead academics? These unpublished works—including manuscripts, correspondence and notes—are known as *Nachlass* (from the German word *Nachlassen*: to leave behind). Multiple works based on the *Nachlass* of the influential philosopher Nietzsche were published after his death; moreover, the philosopher Wittgenstein only published one book during his lifetime, other academics compiled all other works from his *Nachlass*. Many of those spending time working on the *Nachlass* of these dead academics can be emotionally invested in the data, with feelings of responsibility to continue the work of the dead scholar. While some academics are actively involved in creating and preserving their *Nachlass*, other scholars such as Alfred Whitehead wanted all unfinished and unpublished work destroyed after his death because he preferred to encourage young academics to develop their own ideas (Lowe 1982). In educational institutions, we continue to learn from the dead in many ways, but what if your current university course contained an online lecture from a dead professor? In February 2021, a student at Concordia University wanted to email his professor with a question about a lecture he had just watched. Searching for an email address, he found a memorial page for his professor who had died two years earlier (Elks 2021). The university still listed the dead professor on their website and was therefore profiting from the pre-recorded lectures. We are beginning to accept that technology ensures the dead remain active in ways not previously available, but the idea that in the future, these lectures could be technologically—and crucially ideologically—updated in some way raises questions about content rights and how the dead can be manipulated by those who control the data. This news item made me think about how intentional *and* accidental digital afterlives could be seen as unfinished work: the 'narbs' (Mitra 2010) of everyday life suspended and waiting to be interpreted and completed by others. As Arnold et al. note thanatechnology has resulted in a repositioning of the dead within the everyday lives of the living (Arnold et al. 2018).

The concept of a digital *Dasein* suggested here in this chapter mainly focuses on where the accidental digital dead reside. However, when someone *intentionally* creates data to be inherited after they die—thereby showing a desire to be involved in how they are remembered in the future—this intentionality results in a contested and contingent space which forces us to consider *whose Dasein* we are discussing: the digital *Dasein* of the creator or the digital *Dasein* of the bereaved? Because these intentional digital afterlife creators have invested in their online selves, they can be described as 'digital flesh' (Gibson & Carden), thereby existing in their own digital *Dasein* once they are biologically dead. The exponential growth of DAI technologies will undoubtedly result in further tensions: between those who are creating digital afterlives and those who inherit them. This philosophical struggle will only add to the questions surrounding the contested space of the digital *Dasein*. These issues are not resolved within this chapter. Moreover, this chapter will no doubt add more questions to the debate about the ontological status of the dead and the contested space of the digital *Dasein*. However, I am hoping that by introducing the notion of a digital *Dasein* I am providing a starting point for further research and debate. Having grappled with some of the philosophical aspects of digital afterlives, we now turn back to the research finding; this time the findings from those purposefully creating digital afterlives with the knowledge they may be inherited by others after their biological deaths—the DCs.

## References


Arnold, Michael, Martin Gibbs, Tamara Kohn, James Meese, and Bjorn Nansen. 2018. *Death and Digital Media*. Routledge.

Bassett, Debra. 2015. Who Wants to Live Forever? Living, Dying and Grieving in Our Digital Society. *Social Sciences* 4 (4): 1127–1139. https://doi.org/10.3390/socsci4041127.

Bell, G. 2004. Intimate Computing? *IEEE Internet Computing* 8 (6): 91–93.

Boellstorff, T. 2014. Rethinking Digital Anthropology. In *Digital Anthropology*, ed. H. Horst and D. Miller 3rd, 39–60. London: Bloomsbury Academic.

Bollmer, Grant David. 2013. Millions Now Living Will Never Die: Cultural Anxieties About the Afterlife of Information. *The Information Society* 29 (3): 142–151. https://doi.org/10.1080/01972243.2013.777297.

Braman, James, Giovanni Vincenti, Alfreda Dudley, Yuanqiong Wang, Karen Rodgers, and Ursula Thomas. 2013. Teaching about the Impacts of Social Networks: An End of Life Perspective. *Online Communities & Social Computing (9783642393709)* January: 240.

Briggs, Pam, and Lisa Thomas. 2014. The Social Value of Digital Ghosts. *Moreman and Lewis*: 125–141.

Brubaker, Jed R., and Janet Vertesi. 2010. Death and the Social Network. In *CHI 2010 Workshop HCI at the End of Life*, 1–4.

Brubaker, Jed R., Gillian R. Hayes, and Paul Dourish. 2013. Beyond the Grave: Facebook as a Site for the Expansion of Death and Mourning. *The Information Society* 29 (3): 152–163. https://doi.org/10.1080/01972243.2013.777300.

Cann, Candi K. 2014. *Virtual Afterlives: Grieving the Dead in the Twenty-First Century*. University Press of Kentucky.

Christensen, Dorthe Refslund, and Kjetil Sandvik. 2015. Death Ends a Life *Not* a Relationship: Timework and Ritualizations at Mindet.Dk. *New Review of Hypermedia and Multimedia* 21 (1–2): 57–71. https://doi.org/10.1080/13614568.2014.983561.

Clinton, P. 2004. Review: 'Polar Express' a Creepy Ride. 2004. http://edition.cnn.com/2004/SHOWBIZ/Movies/11/10/review.polar.express/

Elks, Sonia. 2021. Analysis: Class Led by Dead Professor Spotlights COVID-Era Content Rights. *Reuters*, February 5. https://www.reuters.com/article/us-global-tech-rights-analysis-trfn-idUSKBN2A521B. Accessed 10 Feb.

Floridi, Luciano. 2015. The Onlife Manifesto. In *Being Human in a Hyperconnected Era*, ed. Luciano Floridi. Springer International Publishing.

Freud, Sigmund. 2004. The Uncanny. In *Fantastic Literature: A Critical Reader*. Westport: Praeger.

Gibson, Margaret. 2017. Grievable Lives: Avatars, Memorials, and Family 'Plots' in Second Life. *Mortality* 22 (3): 224–239. https://doi.org/10.1080/13576275.2016.1263941.

———. 2019. Death in *Second Life*: Lost and Missing Lives. In *Residues of Death: Disposal Refigured*, ed. Tamara Kohn, Martin Gibbs, Bjorn Nansen, and Luke van Ryn, 153–168. London: Routledge.

Gibson, Margaret, and Clarissa Carden. 2018. Living and Dying in a Virtual World. *Switzerland: Palgrave Macmillan Memory Studies*.

Gustavsson, Anders. 2011. *Cultural Studies on Death and Dying in Scandinavia*. Novus Forlag.

Hallam, E., and J. Hockey. 2001. *Death, Memory, and Material Culture*. New York: Berg.

Hallam, E., J. Hockey, and G. Howarth. 1999. *Beyond the Body: Death and Social Identity*. London: Routledge.

Heidegger, M. 1962. *Being and Time*. New York: Harper & Row.

Hockey, J., C. Komaromy, and K. Woodthorpe. 2010. *The Matter of Death: Space, Place and Materiality*. London: Palgrave Macmillan.

Kasket, Elaine. 2011. Being-Towards-Death in the Digital Age Presentation, 19 November, 2011 Society for Existential Analysis Annual Conference, 1–20.

———. 2012. Continuing Bonds in the Age of Social Networking: Facebook as a Modern-Day Medium. *Bereavement Care* 31 (2): 62–69. https://doi.org/10.1080/02682621.2012.710493.

Kim, Joohan. 2017. Phenomenology of Digital-Being Author(s): Joohan Kim Source: Human Studies. 24, (1/2), Intertexts: Philosophy, Literature and the Human Sciences in Korea (2001), 87–111. Published by: Springer Stable URL: Http://Www.Jstor.Org/S 24 (1): 87–111.

Lagerkvist, Amanda. 2017. Existential Media: Toward a Theorization of Digital Thrownness. *New Media & Society* 19 (1): 96–110.

Lowe, Victor. 1982. ANW: A Biographical Perspective. *Process Studies*: 137–147.

Lupton, D. 2015. *Digital Sociology*. 1st ed. New York: Routledge.

Marwick, Alice, and Nicole B. Ellison. 2012. 'There Isn't Wifi in Heaven!' Negotiating Visibility on Facebook Memorial Pages. *Journal of Broadcasting & Electronic Media* 56 (3): 378–400.

Massimi, Michael. 2012. Thanatosensitively Designed Technologies for Bereavement Support.

Meese, James, Bjorn Nansen, Tamara Kohn, Michael Arnold, and Martin Gibbs. 2015. Posthumous Personhood and the Affordances of Digital Media. *Mortality* 20 (4): 408–420. https://doi.org/10.1080/13576275.2015.1083724.

Mitra, Ananda. 2010. Creating a Presence on Social Networks Via Narbs. *Global Media Journal* 9 (16): 22–40.

Mori, M. 2012. The Uncanny Valley. Trans K. F. MacDoram and N. Kageki. *Robotics and Automation Magazine* 19 (2): 98–100.

Nansen, Bjorn, Michael Arnold, Martin Gibbs, and Tamara Kohn. 2014. The Restless Dead in the Digital Cemetery. *Digital Death: Mortality and Beyond in the Online Age*: 111–124.

Odom, William, Richard Harper, Abigail Sellen, David Kirk, and Richard Banks. 2010. Passing on & Putting to Rest. In *Proceedings of the 28th International Conference on Human Factors in Computing Systems—CHI '10*, 1831. https://doi.org/10.1145/1753326.1753601.

Pennington, Natalie. 2013. You Don't De-Friend the Dead: An Analysis of Grief Communication by College Students Through Facebook Profiles. *Death Studies* 37 (July 2015): 617–635. https://doi.org/10.1080/07481187.2012.673536.

Sample, Ian. 2011. Stephen Hawking: 'There Is No Heaven; It's a Fairy Story.' *The Guardian (Online)*. 2011. www.theguardian.com/science/2011/may/15/stephen-hawking-interview-there-is-no-heaven

Sherlock, Alexandra. 2013. Larger Than Life: Digital Resurrection and the Re-Enchantment of Society. *The Information Society* 29 (3): 164–176. https://doi.org/10.1080/01972243.2013.777302.

Stokes, Patrick. 2015. Deletion as Second Death: The Moral Status of Digital Remains. *Ethics and Information Technology* 17 (4): 237–248. https://doi.org/10.1007/s10676-015-9379-4.

———. 2021. *Digital Souls: A Philosophy of Online Death*. Bloomsbury Academic.

Walter, Tony. 2015a. The Dead Who Become Angels: Bereavement and Vernacular Religion. *OMEGA-Journal of Death and Dying*. https://doi.org/10.1177/0030222815575697.

———. 2015b. Communication Media and the Dead: From the Stone Age to Facebook. *Mortality: Promoting the Interdisciplinary Study of Death and Dying* 6275 (May): 1–33. https://doi.org/10.1080/13576275.2014.993598.

Walter, Tony, Rachid Hourizi, Wendy Moncur, and Stacey Pitsillides. 2012. Does the Internet Change How We Die and Mourn? Overview and Analysis. *Omega: The Journal of Death and Dying* 64 (4): 275–302. https://doi.org/10.2190/OM.64.4.a.

CHAPTER 4

# From Digital Footprints to the Ultimate Selfie: The Experiences and Motivations of Digital Creators

*The first condition for immortality is death.* (Stanislaw Jerzy Lec)

Many of us are digital creators: in an age of ubiquitous social networks, and the ever-growing phenomena of self-quantification technologies (including wearable health, fitness and sleep tracking devices), enabled by the Internet, we create memories, messages and other data that others could inherit after we die. However, in this chapter, I focus on the research findings from individuals who are *purposefully* and *intentionally* creating digital memories or messages to be inherited by others after their deaths, along with participants who are using thanatechnology to memorialise the dead. One of the aims of this research was to gain an in-depth understanding of the lived experiences of DCs in an effort to understand this creative process, and to explore their hopes and expectations for how these memories and messages may be experienced and used in the future. I start this chapter by looking at some of the dilemmas faced by virtual ethnographic researchers, before introducing the four distinctive subcategories for the DC group of participants.



D. J. Bassett, *The Creation and Inheritance of Digital Afterlives*, Palgrave Studies in the Future of Humanity and its Successors,
https://doi.org/10.1007/978-3-030-91684-8_4

## Research Lurking: The Public/Private Dichotomy of Social Media Platforms

As mentioned in Chaps. 1 and 2, I used Twitter to recruit research participants and data collection. Although Twitter challenges the traditional notion of what Marwick (2014) describes as a 'bounded' field site, I felt it appropriate to use social media platforms to research social media platforms. However, I needed to consider whether it was necessary to ask permission to use tweets, thanablogs and YouTube comments considering their public/private nature. Others have discussed this issue (see Lange 2007; Qian and Scott 2007; Viégas 2005); however, using tweets about dying and bereavement needed further careful considerations due to the personal and emotional content of the data being created and used. All participants are unique, and expectations of what is public and what is private may vary between people, and it is reasonable to assume that most social media platform users would not be aware that researchers are part of the audience (Moreno et al. 2008). I did consider unobtrusive methods for the research—such as examining tweets without seeking approval from the authors as they were in the public domain—however, 'lurking' is considered a covert research method (Hine 2003) and something I wanted to avoid. There are ongoing debates around the public/private nature of Internet research (see Markham and Buchanan 2012) and special consideration needed to be given to Cupit's warning of being perceived as an 'ambulance chaser' (2012). Following careful consideration of these debates, I chose an overt role as a researcher, whereby I disclosed my research topic and university affiliations to all potential interview participants. Because the research focused on data from terminally ill or bereaved people, I made the decision that no form of ethical 'lurking' would be appropriate; therefore, I sought consent to use *all* data in whatever form it took including publicly available tweets and thanablogs. During the research, I found several YouTube videos created as goodbye messages from people who knew they were dying. These were useful to the research and provided data for me to explore the motives of DCs, and the comments underneath the videos allowed me to see how people were reacting to these posthumous messages. From the comments, it was possible to identify that many of the respondents to these videos were tertiary mourners—that is, they did not know the deceased—which have been described by DeGroot (2014) as 'emotional rubberneckers'. However, I only captured and coded comments where it was apparent that the person leaving the comments knew the deceased. These transcripts provided vital data from a 'missing' group of digital inheritors: those who may have been reluctant to participate in in-depth interviews.

In total, I interviewed 17 DC participants who were using various platforms on the Internet to create a variety of digital memories and messages which they hoped would outlive their biological existence. During the early stages of the recruitment process—and in the early days of data collection—I considered all DC participants together under one category. However, it became clear that there were fundamental differences in the types of digital memories and messages these DC participants were creating, and more importantly, their motivations differed. Therefore, I began to realise that these DC participants needed further categorisation. Some DC participants who were terminally ill were creating memories and messages which they hoped would continue to exist after their deaths; other participants—who were not terminally ill—were also creating personal messages to be delivered after their deaths; some participants were creating blogs and websites following the death of a loved one, while others were creating avatars and participating in life-logging in an effort to intentionally create digital afterlives. It became apparent that although these 17 participants were DCs, to treat them as one group without subcategorisation would not result in any clear understanding of their motives and hopes for this digital data. Therefore, I developed four subcategories that described the 17 DC participants:

**Thanabloggers**
As discussed in Chap. 1, Sofka's term thanabloggers (2012) describes those who use the Internet to blog about their personal experiences of death and dying. Those creating thanablogs are aware that their digital data will outlive them. I recruited a total of five female thanabloggers (interestingly, all efforts to find and recruit male thanabloggers failed).

**Creators of posthumous messages**
This subgroup consists of participants who are not terminally ill but have nevertheless decided to record messages to be delivered to their loved ones following their deaths. Some of these participants have created messages to be delivered on specific future dates. In contrast to the thanablogging DC participants, both of the participants who were creating posthumous messages were male.

**Grieving digital creators**

The social norms and rituals around funerals are changing, and funeral selfies and live-streamed funerals are just two examples of the effect technology has on mourning practices. The creators of funeral films, thanabots and grief blogs are using the Internet as a creative space where the process of creation can be comforting. Funerals of the famous and infamous have always been considered newsworthy; however, one—previously unknown—family made the conscious decision to produce, edit and publish a film of their beloved son's funeral. Josh tragically died in 2011, aged 22, following a road accident in Vietnam. Following his death, his parents Jane Harris and Jimmy Edmonds filmed and published a film of Josh's funeral. During the film, they explain how they wanted to speak openly—and encourage others to speak openly—about the loss of a child. Inevitably, grieving digital creators overlap with the DI group of participants: once they have created the digital afterlife of their loved one, they then inherit their own creation. I recruited five DC participants—three female and two male.

**Early adopters**

This subcategory of DCs includes life-logging participants who are creating Ancestor Veneration Avatars (AVAs), or creating avatars of themselves on the platforms discussed in Chap. 2. AVAs (Bainbridge 2013) are avatars embedded within virtual worlds and based on historical biographies of a deceased person. I was successful in recruiting a participant who had embedded some of his deceased relatives into virtual worlds, along with four participants who have created avatars of themselves in an effort to create digital doppelgängers, which they hope will remain active after they die.

## Digital Scholarship and Digital Sociology

Rather than following a linear process, with the iterative methodology of grounded theory, reviewing existing literature, data collection, coding, identifying themes, identifying relationships and drawing conclusions to produce conceptual theories—all happen simultaneously. Throughout the research, I engaged in what Weller (2011) describes as 'digital scholarship'. As the research aims were to explore how others experience certain phenomena within the digital realm, digital scholarship seemed apt. Digital sociology focuses on how people use digital technologies and the data they produce; therefore, I adopted what Lupton (2015) describes as a 'professional digital practice'. I quickly realised that any attempt to understand and explore our digital society would require me to fully immerse myself within that society. However, I had never been comfortable having an online presence and preferred a Google search on my name to deliver no results—especially no images. I would ask friends to remove photographs of me from their Facebook pages as I valued my privacy. It was therefore difficult for me to consider that in order to research social networking sites I needed to open profiles with Facebook, Twitter and LinkedIn. As a researcher, I adopted what Lupton (2015) described as a 'professional digital practice' to build networks and create online profiles in order to recruit participants and publicise the research. However, going from no online presence to having these profiles and a website was something I found difficult to get used to, but in order to work within digital scholarship, it became necessary to be 'searchable' to potential participants and other scholars. Social media differ from other media as they have specific characteristics: persistence, visibility, spreadability and searchability (Boyd 2014). These characteristics offer researchers opportunities due to the lack of established gatekeepers forming a barrier to ethnographic research (Carrigan 2016). I found many positive reasons for using social media to recruit and collect data, but also using Twitter to communicate with other academics proved invaluable. This supportive online community supplemented the face-to-face discussions at conferences and networking events by offering what Carrigan (2016) describes as further opportunities for fruitful collaboration and clarifications long after leaving these events. Interestingly, through blogs, websites, Twitter activities, publications, online interviews and press mentions I became a digital creator—something I had never intended to be.

## The Motivations of Digital Creators

Understanding the motivations behind people creating digital memories and messages with the intentionality that others will inherit them following their deaths was vital to the research. Following the data collection, transcription and coding five themes which were consistent across all 17 DC interviews: (1) Benevolence (2) Therapy (3) Continuing bonds (4) Technophilia and (5) The quest for digital immortality. In the next section of this chapter, I explain these five themes in greater detail,

## Benevolence

The word benevolence reflects the prevalence of kindness and well-meaning throughout many of the DC interviews. Some participants commented on the importance of creating digital memories and messages that could help and educate others in similar situations, while others discussed how their digital messages might help health professionals. Five codes (helping others; inform/educate others; truth-telling; improve future care and raising awareness) underpinned this theme of benevolence and were used in the early stages of analysis when it became evident that participants were aware that the digital data they were creating could improve the lives of others.

Having recruited five thanablogging, I began to see this emerging theme of benevolence. Jan, Tracey and Mary spoke passionately throughout their interviews about using social media as a crucial advocacy tool and their determination and desire to help others with potentially terminal illnesses. The following quotation—which is typical of many of the thanablogging participants—illustrates this desire to help and inform others:

> *Writing it down it was a very good way of me concentrating my thoughts and working out what I was actually feeling [...] a patient treated in the same hospital as me had written a blog which doesn't really give a good message about breast cancer [...] but I then suddenly realised a lot of blogs that patients write are quite scary to read, and I wanted to make things seem a little bit more calm and sensible, and less daunting.* (Jan)

The following two extracts from the interview with Tracey demonstrate how the ability to create digitally mediated messages—that can be accessed by other people—makes thanablogging DCs feel their experience of illness can be helpful to others in similar situations:

> *You know to give people information and give them power to help them in their decisions, you know so that it makes it easier for them and I think that's where I started from.* (Tracey)

> *If I can pass any kind of information on to people then that's great, there is other people who have contacted me through the advocacy and the blogging, that again I have been able to share information of theirs so again it's passing that information along.* (Tracey)

The theme of benevolence is also found in similar research: Taylor and Pagliari (2018) analysed the tweets of thanablogger Kate Granger during her final six months; they found their primary dimension 'physical symptoms' received 270 tweets during this time. However, I was interested in a category that was excluded from their results—but which was useful to the research—their findings showed that Kate Granger tweeted about campaigns including #hellomynameis and fundraising activities, a total of 388 times during her final six months. These findings echoed what I was learning from the thanablogging participants: they predominantly used Twitter and thanablogs to advocate for change and help to others.

Jan told me how she found the blogs of other terminally ill people extremely useful and how they affected how she felt about her own illness:

> *It's an accessible morbid curiosity that I wish I don't have access to, and for me when I found out my cancer had spread I spent the night on Google. I know everything, so I was desperately looking at forums that mentioned it had spread thinking 'oh, my god! she's dead in a year, it could be me'. Just this insane thirst for any knowledge. And I don't know whether you are looking for reassurance or reinforcement or refusal, but it is quite scary and, depending on which blog you read, and how it's written, can dramatically change how you feel about things.* (Jan)

In addition to informing and advising other cancer patients, Jan's motivations included the hope that healthcare professionals could benefit from reading thanablogs. Jan explained how she believed that patient care might improve as a result of these frank and honest blogs:

> *I thought if I can write a blog it might help doctors realise that you are not just chemotherapy, it's not just a biopsy, to maybe help them improve patient care in the future.* (Jan)

> *Blogging is a big part of my life because it does share that information and it's been a positive thing for me, certainly not been a negative thing.* (Tracey)

Alongside suggestions that informed and educated readers about their experiences with cancer, two participants also told me how blogs enable them to be frank with readers about the reality of their day-to-day lives. Both Tracy and Jan explained how the public nature of social media allows thanabloggers to include 'truth telling' in their blogs and why they felt this was an essential part of these online digitally mediated messages:

> *But that is why, because people do need to see that. The thing is, it's not pink and it's not fluffy.* (Tracey)

Here, Tracey was referring to The Breast Cancer Campaign, and Breast Cancer Now's using the colour pink and bows to highlight and fundraise for people living with breast cancer. In contrast to these campaigns, the thanablogging participants in this study told me how they were attempting to paint a more realistic picture of their experiences. Jan talked about how she felt about television personalities who discuss their potential terminal illnesses on television programmes and how she did not think these programmes reflected the experiences of many cancer patients:

> *It's an artificial … we can't all look like you, in their house looking fabulous having chemo, others that are naked on the loo trying to have a crap, looking like shit.* (Jan)

However, this motivation of 'truth telling' also highlighted a recurrent concern for the participants: in that writing their digitally mediated blogs—and reading blogs written by others—made them contemplate the reality of their own mortality:

> *I want to talk about my funeral and actually accept that it might happen. Whereas before you might keep your head in the sand.* (Jan)

> *For me as a cancer patient it makes me face up to my own mortality.* (Jan)

Bringing discussions about death and dying out of the shadows and helping others seem to be a key motivation for thanabloggers. However, some participants expressed the view that there is a line that should not be crossed on thanablogs: one participant explained how she found some digital memories created by other thanabloggers to be upsetting rather than helpful:

> *The thing that really upset me … erm … was the picture that was tweeted from the hospice bed … and that made me … I recoiled in horror.* (Jan)

It also prompted Jan to discuss her wishes with her husband:

*And I told my husband if I ever get that ill I don't want you to put a photo of me looking like that on social media because that's not how I want people to remember me … or see me.* (Jan)

This unease was echoed in the interview with Tracey, who also spoke about not wanting to see deathbed scenes on social media. She described how she felt when another thanablogger did this:

*She actually did post one of the pictures of her dying … and it was horrendous … and it will stay with me forever … you know it's one of them things that you just think oh … awful.* (Tracey)

Tracey described how distressing it was to see post-death photographs published on social media sites of urns and coffins. She went on to explain how uncomfortable this made her feel, adding that she did not want this to happen if she died:

*What I didn't like what I saw the other day was a picture of the urn … now I thought I found that a little bit OTT and I actually messaged another friend who doesn't blog and I said to her if my husband does that when I'm dead, I said you're allowed to come and kick his backside.*

Thanablogs will be an ongoing and important area of study for researchers. The immediacy of blogs and tweets, and their ability to reach others in similar circumstances, help to highlight the experiences of the terminally ill (Taylor and Pagliari 2018). Throughout the interviews with the thanablogging participant, it was evident they found comfort in both creating and reading them.

Grieving DC bloggers also discussed this idea of helping others through blogging on the Internet. Grief bloggers differ from thanabloggers in that they are blogging about the loss of a loved one rather than blogging about their own terminal illness—or contemplating their own death. Two of the four grieving DC participants expressed the importance of blogging to help others whilst also dealing with their own grief. During the interview with Amy—whose sister had died—she told me how she was motivated to use SNS to talk about her experiences:

*I actually want to help people who are either facing similar situations or have experienced loss.* (Amy)

Amy went on to explain how, for her, it was not only important to share her own experiences on SNS, but she wanted to be there to listen to other people's stories in an effort to help them with their grief:

> *There was also the aspect of actually counselling quite a lot of people I spend a lot of time in the first few months … checking messages, writing messages.* (Amy)

> *When they have similar situations and that I can offer maybe some advice and you can … be there for people when they want that.* (Amy)

During his interview, Ivor—whose daughter had died—also talked about supporting others through blogging. Ivor had written a blog during his daughter's illness and explained how it felt like it had become a daily drama or 'people's EastEnders'. Following his daughter's death, he used the blog to inform people about the funeral arrangements and then wrote a final blog where he thanked people who had supported the family through their daughter's illness. However, following his 'final' post, he received requests for him to continue with his blog:

> *and I almost felt, you know, how, you know, when you're grieving you end up supporting other people at the time when you need support. And I almost felt I had to carry on, I couldn't … ya know … I was letting these people down if I didn't.* (Ivor)

I was seeing a tension between the needs of the blogger and the perceived needs of the reader. Nevertheless, while this tension exists, helping others seemed to make a difference to some participants who were facing death or grieving for their loved ones.

Once I had created the early adopter DCs category, I wrongly assumed most of the coding would be of a narcissistic or selfish nature as I was unsure how anyone else could benefit from this type of digital creation. However, this was not the case: the early adopters coded heavily under the theme of benevolence, with three out of four participants discussing how they wanted to create data that could be useful to researchers or anthropologists in the future.

Paul had created an avatar on the LifeNaut platform, and I wanted to understand his motivations for spending time on the creation and ongoing training of his avatar. He encouraged me to create an avatar of myself to 'allow researchers interactive access' to my work. He explained how it would allow researchers to connect with me on a 'virtual level' and—more importantly, he told me—at a time when I 'might not be available'. He added:

> *If a scientist wanted to interact with a fellow scientist who may have passed on, it would give them the opportunity to virtually work with that person to a degree, depending on how detailed the MindFile was and how well the Avatar can converse about the data.* (Paul)

Tom—another early adopter DC participant—was a life-logger who actively collected and curated data from all aspects of his daily life. During his interview, he explained his motivations for this digital creation:

> *I guess it's more of a that, I want to have content there, so that if people want to go back and look at what happened at these various environments or stuff that ... those are available mostly for sort of management or historical, ya know, studies. I have been a part of all of these organisations and so it's more ... here's the part I played in that. So that's what I kind of see the value.* (Tom)

Paul described the potential usefulness of his avatar creation to future generations who may seek to understand what life is like in the twenty-first century:

> *Let me record my life in this way, and if it is a benefit to someone far off in the future, then I'll have done a good thing and perhaps helped someone better understand themselves by understanding me and my generation.* (Paul)

The usefulness of this type of digital creation was similarly described in an interview with Russell, who explained his motivations for creating an avatar of himself:

> *I think we are making history or contributing to that part of history as we go along so I think I might leave it open just for, ya know, future investigation, ya know, like students like yourself who might wanna go back and investigate what was going on back in 2016.* (Russell)

Benevolence was the most heavily coded theme in the interviews with the DC participants. Yet not all the subcategories of this group of DCs coded under this theme: neither of the two interviews with the posthumous message creators contained any suggestion of benevolence being a motivation for their digital creation. However, I still wanted to understand if leaving posthumous digital messages could ever be seen as a benevolent act; searching through the YouTube channel, I found a video which had been recorded and published by a terminally ill creator who was leaving a message to her son. It had been viewed 1,272,301 times and had 3618 comments (on 14 December 2020). The following quotation is from the transcript:

> *I hope you remember me and I hope this time-capsule helps. When you are watching this, know that I have moved on and that I am in a better place because Mom's not suffering anymore. But I miss you so much and it hurts to know that you're gonna go through all of this without your mom.*

This extract demonstrates how leaving digital posthumous messages can be seen as a benevolent act in some circumstances.

## Therapy

This second theme reflects what participants told me during interviews about how they used digital messages created on SNS as a form of therapy. This theme was prevalent through many of the participant interviews, and there were three codes that underpinned the theme: self-understanding, self-help/protection, creative/productive.

One of the DI participants, Grace, told me how her husband had blogged throughout his illness. She spoke of what this had meant for her husband and told me, 'you know it gave meaning to the process of dying'. In the interview extract below, Jan—a thanablogger—describes how for her blogging and tweeting were a good way of communicating bad news to a large audience, rather than repeating it multiple times. Jan explained how constantly repeating bad news verbally rather than digitally could be both mentally and physically exhausting.

> *but by reading the blog they felt they were more in touch with me. They were upset I couldn't tell them everything on the phone, but often during chemo you can't … or … it's a lot to say. But they just felt … we know what's going on … we're in touch.* (Jan)

Jan expressed how this was a part of her 'self-preservation' as she was 'protecting' herself by doing this. This comment from Jan resonated with a blog post from another thanablogging participant; Tina's post also described the need to use SNS to communicate bad news, rather than sending individual messages to a wide number of people:

> *This is the blog post I hoped I'd never write. A lot of people who aren't yet aware of my situation will be shocked and upset by it. There are so many people I'd like to break the news to on an individual basis but that's just not possible. Whoever you are, thank you for reading.* (Tina)

Similarly, Ivor told me how he also found blogging a useful way to communicate with family and friends about the deterioration of his daughter's health and then—following the death of his daughter—communicating to people about the funeral arrangements. He preferred this method of digital communication rather than trying to communicate to people individually, which he explained can be time-consuming and mentally draining. Ivor explained:

> *We didn't have to go through the story at the end of every day ... Then it became our tool for telling people what was happening and then at the ... when she died, we used social media to tell people about the funeral.* (Ivor)

Occasionally this theme of therapy was implied in interviews where participants commented on the importance of *not* creating or reading grief or thanablogs. Jan explained how sometimes she made the decision not to visit these SNS as a form of self-preservation:

> *and there is the grim curiosity and I think ... no ... that's enough put it away for a month, don't go there. And you kind of dip in and out and feed this demon inside you, does that make sense?* (Jan)

Self-preservation is also important to those dealing with the death of a loved one and blogging about grief. Ivor explained how he needed to moderate the frequency of blog posts following the death of his daughter:

> *I'd look at those and then ... I'd turn away from it and then I started gradually dripping myself back in the new year and I think there is something about ... this ... about this sort of self-protection.* (Ivor)

Interestingly, Paul—one of the early adopters—also coded under this theme of therapy. He explained how the process of creating his digital avatar was not just a process of self-capture. Rather he described how it had helped him understand himself:

> *I've discovered that, in a way, creating my Avatar has helped me better understand who I am, as I now see more clearly how the pieces of my puzzle fit together and make up who I am today.* (Paul)

## Continuing Bonds

> *I can either post a card in a letterbox to her in Heaven or I can do something on this magical Internet Facebook stuff that gives us all this bullshit.* (Pam)

When someone dies, the desire to continue a relationship with the deceased is easily understandable. As we saw in Chap. 1, a plethora of literature addresses how the Internet enables new ways of preserving bonds with the dead. Through the voices of the participants who are using digital memories—and in the messages they inherited following the death of a loved one—it is therefore hardly surprising that the findings of this research would reveal a theme of continuing bonds. However, this theme was also evident and coded heavily in interviews with those doing the creating. Within the data, this umbrella term of continuing bonds included the codes of keeping memories alive, reminding others of their grief and staying connected to the dead. During the interview with Jess—a grieving DC participant—I asked if she could explain how important it was for her to create digital memories to remember and memorialise her son:

> *When somebody dies you think that's … if you like, the end of their life. But until you experience it at first hand you don't quite realise that death brings a lot of new experiences, and a lot of new learning, and that … particularly with a child … that child just because they're dead doesn't mean to say they are not part of your life.* (Jess)

The theme of continuing bonds was even more explicit in the following extract from the interview with Jess:

> *but it became more than that, because really what we wanted to then, was to continue that relationship on … continuing the relationship with our son.* (Jess)

Evidence that thanatechnology facilitates ongoing relationships with the dead was similarly reinforced in other DC interviews:

> *Yes, like I just pull out my phone, and I open the Facebook messenger App … and then it's just as if you were sending text messages back and forth with a friend … but you're doing it with, with the Dadbot.* (Noah)

In the interview, I asked Noah how much he has changed or added to the Dadbot since its initial creation:

> *I can always add new little anecdotes for him to share, or new varieties of ways for him to express whatever he already talked about.* (Noah)

Noah explains that he can work on the listening side of the Dadbot and improve how the bot reacts to what he or someone else says to it. However, he explained how he felt there was a creative 'line' which he did not want to cross:

> *I was always wanting it to be … better and better and better … you know with more … knowledge and … better able to respond … and more of my father's signature vocabulary … so I always wanted to improve it … but then at the same time there is this kind of weird … I know there is a line … out there where … that it would somehow get … too good.* (Noah)

I asked him if he could expand on this for me:

> *becoming almost like my Dad but it wasn't really him and … I've struggled with … been asked … you know where is that line … like what would … what if you could do him as an animation and what if you could have a synthesis of his … real voice … and … I knew … I know that's a line out there and when I think about an animated version of him for instance that just gives me the creeps … personally.* (Noah)

It seemed that any form of reanimation still provoked feelings of the uncanny valley—even for those doing the digital creating. The research finding from the DC participants highlighted the links between thanatechnology and the theory of continuing bonds. However, although these findings resonated with other existing research (see Bouc et al. 2016; Degroot 2012; Getty et al. 2011; Kasket 2012; Pennington 2013; Walter et al. 2012), significantly they revealed a more nuanced aspect of the theme of continuing bonds: the data illustrated how some participants wanted to make sure—through digital memory and message creation—that their deceased loved not only remained a part of *their* lives but also remained in the lives of others too:

*I always did because it keeps her real. To start with I wanted to make, well not make sure, I didn't want people to think she wasn't there anymore.* (Pam)

Similarly, this motivation was reinforced by Jess:

*You know he will live on through these records, and we like that thought, so that you know we're not just carrying him on, everyone who accesses this digital record of him can learn about him if they want to.* (Jess)

In a similar vein, Jess described the need for things to be put in the 'correct' chronological order. During the interview, she explained that although her son had died before her, it was important to her that something would remain of him after she died:

*We very much want them to be a permanent archive or a record of our son's life for anyone and everyone to be able to access as they want, because they endure, and they live on after we've died.* (Jess)

Two of the DC participants explained how as time passed, they felt it necessary to remind people they were still grieving, and social media gave them a platform to achieve this. Extracts from the interviews with Pam and Ivor (presented below) illustrate these poignant and honest feelings:

*It's quite a selfish thing actually, maybe it's just to make sure that people know that your living through shit.* (Pam)

She went on to explain:

*It's just to make everybody else aware that I am still thinking of her.* (Pam)

Similarly, during the interview with Ivor, he spoke of his reasons for still creating posts about his daughter on Facebook almost ten years after her death:

*It makes me think … I'm not gonna let you forget her.* (Ivor)

He hesitated, and then added

*But then there is also this thing around, do you know what, this will make you feel how it makes me feel.* (Ivor)

In the interview with Jess, she described how she wanted to create digital memories of her son because she realised that she would *never* find closure:

> *I suppose we didn't realise at the time we wouldn't find any closure but we would learn to live with the loss.* (Jess)

This realisation was similarly described in the interview with Ivor, where he discussed how creating digital memories enabled him to manage his grief:

> *I don't say that I cope, but I manage my grief. Well, you know, time doesn't heal but I think what time does do is give you the ability to manage your grief.* (Ivor)

We will be taking an in-depth look at grief theories in Chap. 6; however, it is interesting to note how the—now largely debunked—theory of moving on and letting go was utterly absent from any participant interviews. Social media provide the bereaved with tools to create digital memories and messages that can be used to facilitate—in a unique way—the creation of continuing bonds to memorialise, remember and importantly *remind* people about the dead.

## Technophilia

Following the initial coding of the DC participants, it became apparent that there were four transcripts that I had found extremely interesting but had almost no coding. Following further analysis, I identified that they all had one thing in common: they all loved the technological capabilities offered by the digital afterlife platforms. Technophilia is the love and enthusiasm for new technologies, and this theme highlighted the motivations of DCs who desired to create digital messages and memories which they hoped would be stored and readily available for future generations by creating what Jacqueline Ford Morie described as the ultimate selfie (2014).

To understand the motives and experiences of those creating and training their avatars, in early 2017, I interviewed Paul:

> *I'm still working on myself ... literally. Every time I interact with my avatar, it becomes more like who I am today.* (Paul)

I wanted to understand if Paul felt that this was different from physical self-capture, such as keeping a diary:

> *Sure, a person can write an autobiography, and fill volumes of books with so much of their life's many adventures. However, the person reading each book can only do just that, read what is written, but not interact with the author.* (Paul)

Paul described how he made certain choices when he created his avatar:

> *I like that I was able to upload a picture of myself that is brought to life and animated by the program. I chose a photo of me as I was when I was 50 years old. I plan to inset a younger version of myself, and of course more recent photos of me, so that the user can choose to interact with my Avatar in the form that they would be most comfortable with.* (Paul)

I asked Paul if he thought other people should take up the opportunity to create avatars:

> *For so many people who have not taken the time to create an avatar of their own, they are missing out on an incredible opportunity to leave behind a virtual version of themselves that can be as fun to interact with as they were. It all depends on how honest they are about themselves, and how much detail they are willing to put into it.* (Paul)

Other participants were also intentionally creating digital afterlives; Life-logger Tom self-captured every aspect of his daily life by wearing a camera around his neck, which took a photograph every 30 seconds. He explained that he felt that although his digital creation was extreme, at the same time, everyone else is accidentally self-capturing their lives:

> *It would seem like you have everybody on Facebook because a Facebook timeline in essence is everybody is life logging whether they like it or not because they have … I don't know whether people consider their life logging or not, but it's certainly, that's what it is.* (Tom)

This idea of everyone creating digital memories and messages is compelling; with wearable technologies that are able to monitor users' heartbeats, steps and exercise, people become 'quantifiable' to many large technology companies. Moreover, as Tom points out, SNS such as

Facebook and Twitter ensure people's daily lives, political opinions, likes and dislikes are recorded, stored and monetised.

> *Facebook hit 2007 and so essentially you now find your life splattered everywhere.* (Tom)

The idea of being able to back up your mind was important to two early adopter DCs:

> *I've long worked on statistical methods for personality capture.* (Ben)

> *the idea of can you back-up a mind and can you … you know … basically save and pass forward information about the human experience.* (Russell)

The importance of digital capture as an aid to memory was also discussed as a key motivation:

> *My motivation is the push to kind of extend my personal … or to have a totally perfect memory.* (Tom)

> *For people suffering from Alzheimer's and dementia, having an avatar would allow them to record so much of who they are and what they've done before they forget completely about it themselves.* (Paul)

The theme of technophilia was particular to the four early adopters interviewed. It demonstrated their love of the nascent digital afterlife technologies, which allow them to digitally capture themselves today, with the explicit hope that people would have the enthusiasm and ability to access their creations in the future.

## The Quest for Digital Immortality

In our digital society, those who use SNS are unintentionally creating a digital afterlife that will likely outlive their biological existence. However, many other people use the Internet—and specifically social media platforms—with the explicit intention of trying to achieve a type of immortality. Not long after I started the research, I attended a funeral where the deceased had written her own eulogy—a carefully airbrushed version of her life. It could be said this is what happens at most funerals; however, writing your own eulogy insinuates that you (a) do not trust anyone else to write a complementary eulogy, or (b) you want to control how you are remembered. The idea of eulogising oneself echoes the motivations behind digital posthumous messages; which can be seen as a digital form of posthumous identity management. Earlier in this book I suggested the term 'digital immortality' was unsuitable when describing the enduring memories and messages enabled by the Internet due to the precarious nature of the service providers. However, here in this section, I need to revert back to using 'immortality' because—for these particular participants—the 'quest for digital immortality' was an active and intentional process. Underpinning this theme were the codes, image control, legacy creation, advise/inform future generations, story sharing and the fear of being forgotten.

Two of the thanablogging participants briefly talked about legacy creation, but only once I had prompted them by asking about it at the end of their interviews:

> *Something that I can have left behind that I have created that will be … I don't know using the word legacy is a bit grand but, erm, if that makes sense.* (Shirley)

> *I do hope that, you know, if when that time comes, that it will hopefully give my children an insight to what I have gone through.* (Tracey)

However, as Tracey's interview demonstrated, even when 'story sharing' and 'legacy' were intimated by some participants, they still reflected the theme of benevolence:

> *To help people get the right information at the time, I think I just wanted to tell my story like everybody else does, you know to give people information and give them power to help them in their decisions, you know so that it makes it easier for them.* (Tracey)

> *But I never think about it as leaving a legacy. I think about what you can do now.* (Tracey)

In a similar vein, only one grieving DC participant interview had coding under the focused code of 'legacy'. After having discussed how precious digital memories and messages left by her sister had become since her sister's death, Amy reflected on her mortality:

*I need to start thinking of kind of my own legacy and the legacy ... because, you know if something happens to me no one has backup.* (Amy)

The desire to advise and inform future generations was explicit within the data from the posthumous message creators, and at the time of the interview with Gary, he had recorded ten messages to be delivered after he dies. He explained his motivations:

*Yes, we live on a digital age – our digital identity will remain behind forever, why not to make it something that will reflect the real us for generations to come?* (Gary)

*you can decide whether you want to have the control and the last words under your erm ... your terms or would you like Google to determine how you will be remembered for generations to come.* (Gary)

Gary used the term 'generations to come' in both of these quotations. Creating a node within NVivo, I wanted to see if others that had coded under 'digital immortality' had mentioned this desire to influence future generations, and they had:

*It may sound weird but that feels good they're gonna hear from me on those birthdays, whether I'm here or not.* (Bob)

I was interested in whether this desire to influence future generations would still be apparent if the posthumous message creator knew they were about to die. So, I transcribed a message left on one of the posthumous message sites where the male digital creator knew he had a terminal illness (he has since died):

*I am hoping for the future generations of the family will get some comfort and some guidance in looking at this message I am sending.*

*Taking on responsibility for my future generations now just to let them know who I am what kind of character I was and my beliefs, and what my contribution was to the world.*

Similar to the findings from the posthumous message creators, Ben, one of the early adopter DCs, also coded under this theme. His desire to advise/inform future generations was evident from his interview and echoed Litton's (1979) idea of symbolic immortality:

> *I tend to think that virtual world avatars are but one of the many ways deceased family and friends can continue to shape reality.* (Ben)

Interestingly, image control and the need to control how others see you after you die were implicitly discussed by all the digital posthumous message creators:

> *I think it's also a mistrust isn't it? It's a mistrust of those who are left behind to remember me.* (Bob)

The need to control how others remember him was similarly described in an interview with Gary:

> *Yes, we live in a digital age, our digital identity will remain behind forever, why not to make it something that will reflect the real us for generations to come?* (Gary)

During the interview with Bob, he described his desire to live on virtually after his biological death through his creation of posthumous messages and his need to ensure his 'personal posterity'. To understand the types of messages Bob had created, I asked him to describe the content of a few of them and when he had timed them for delivery:

> *The 18th and 21st birthdays are not profound, they are just me being myself, and being a bit silly and having fun. Other years are more profound … okay … but 18 and 21 are to be celebrated and not to be serious.* (Bob)

I wanted to know how far into the future Bob had created timed messages for his children to receive:

> *At this point they will get them up to 40. It's just a matter of going in and getting them further but they are covered up to 40 years old.* (Bob)

I asked Bob how he felt knowing that if he died, these messages would be delivered to his children on specific dates:

*It gives me a great deal of pleasure to know that my son who is 8 and my daughter who is 4, they're getting messages from me on their 18th and 20th birthday no matter what, and that feels good to me. It may sound weird by that feels good they're gonna hear from me on those ... birthdays ... whether I'm here or not ... I'm gonna be part of those birthdays ... unless they don't want me to be and they can turn it off but otherwise, yeah, I'm there.* (Bob)

Bob's final comment, in the extract above, about the ability of DIs to choose whether to listen to messages once they are aware they exist, will form a significant part of the discussion around control of inherited timed messages in the next chapter of this book.

Being remembered proved an important motivation for the creation of digital memories and messages; the following quotations from the interviews with Ben and Paul illustrate how the fear of being forgotten manifested through many of the interviews:

*My family has a very long history of developing very intense personal identities, seeking fame but in idiosyncratic ways, and preserving information about family members.* (Ben)

*And no one will be able to speak with you about your many life experiences ever again.* (Paul)

The ability to narrate your own biography can be seen as a form of image control, and importantly a desire to control how you are remembered in the future. In contrast to these 'immortality' participants, the motivations of the early adopters—who are creating avatars and life-logging—were more varied than simply the desire to be remembered. Digital afterlives were important to them, but as one early adopter DC explained, he was not completely convinced that digital immortality is a good thing:

*The reality maybe they've already lived too long anyway, and just do we need them around, do we need any more of me around any more longer than I have been ... I'm not sure.* (Tom)

In contrast to Tom's view of digital immortality, another early adopter looked at the creation of avatars as a way to compete in our digital society:

> *For some people to live longer on this limited Earth, other people must not be born. So I tend to view avatars as allies in competition with people outside the family or other social group.*(Ben)

Interestingly, Ben told me that he had never experienced grief, although most of the close members of his family had died (some in tragic situations). He explained that he did not think he was capable of experiencing grief but that he had spent many hours creating avatars of his ancestors and embedding them within virtual online worlds. I asked Ben if he would like to receive timed digital posthumous messages services from his ancestors and received the following philosophical answer:

> *Sure, fine, as a step toward having an army of revived ancestors doing my bidding as servants. Or, perhaps they will dominate me.* (Ben)

While inevitably some creators are looking to outlive, deny or even conquer death, for others, digital afterlives are about the lives and deaths of others. Therefore, in the next chapter, we will look at the final category of research participants, those who inherit digital afterlives—the DIs.

## References

Bainbridge, William Sims. 2013. Perspectives on Virtual Veneration. *Information Society* 29 (3): 196–202. https://doi.org/10.1080/01972243.2013.777312.

Bouc, Amanda, S. Hye Han, and Natalie Pennington. 2016. 'Why Are They Commenting on His Page?': Using Facebook Profile Pages to Continue Connections with the Deceased. *Computers in Human Behavior* 62: 635–643. https://doi.org/10.1016/j.chb.2016.04.027.

Boyd, D. 2014. *It's Complicated: The Social Lives of Networked Teens.* Yale University Press.

Carrigan, M. 2016. *Social Media for Academics.* Sage.

Cupit, Illene Noppe. 2012. Research in Thanatechology. In *Dying, Death, and Grief in an Online Universe*, 198–214.

Degroot, Jocelyn M. 2012. Maintaining Relational Continuity with the Deceased on Facebook. *OMEGA-Journal of Death and Dying* 65 (3): 195–212.

———. 2014. 'For Whom the Bell Tolls': Emotional Rubbernecking in Facebook Memorial Groups. *Death Studies* 38 (2): 79–84.

Getty, Emily, Jessica Cobb, Meryl Gabeler, Christine Nelson, Ellis Weng, and J.T. Hancock. 2011. I Said Your Name in an Empty Room: Grieving and Continuing Bonds on Facebook. In *Proceedings of the 2011 Annual Conference on Human Factors in Computing Systems (CHI '11)*, 997–1000. https://doi.org/10.1145/1978942.1979091.

Hine, C. 2003. *Virtual Ethnography.* Sage.

Kasket, Elaine. 2012. Continuing Bonds in the Age of Social Networking: Facebook as a Modern-Day Medium. *Bereavement Care* 31 (2): 62–69. https://doi.org/10.1080/02682621.2012.710493.

Lange, Patricia G. 2007. Publicly Private and Privately Public: Social Networking on YouTube. *Journal of Computer-Mediated Communication* 13 (1): 361–380.

Lec, Stanisław Jerzy. 1962. *Unkempt Thoughts.* St. Martin's Press.

Litton, Robert Jay. 1979. *The Broken Connection.* New York: Simon K Schuster.

Lupton, D. 2015. *Digital Sociology.* 1st ed. New York: Routledge.

Markham, Annette, and Elizabeth Buchanan. 2012. Ethical Decision-Making and Internet Research Recommendations from the AoIR Ethics Working Committee. In *Recommendations from the AoIR Ethics Working Committee (Version 2.0)*, 19. Retrieved from www.aoir.org.

Marwick, Alice E. 2014. Ethnographic and Qualitative Research on Twitter. *Twitter and Society*: 109–122.

Moreno, Megan A., Norman C. Fost, and Dimitri A. Christakis. 2008. Research Ethics in the MySpace Era. *Pediatrics* 121 (1): 157–161.

Morie, Jacquelyn Ford. 2014. The 'Ultimate Selfie': Musings on the Future of Our Human Identity. In *Proceedings of the 23rd Annual Conference on Behavior Representation in Modeling & Simulation*, 93–102.

Pennington, Natalie. 2013. You Don't De-Friend the Dead: An Analysis of Grief Communication by College Students Through Facebook Profiles. *Death Studies* 37 (July 2015): 617–635. https://doi.org/10.1080/07481187.2012.673536.

Qian, Hua, and Craig R. Scott. 2007. Anonymity and Self-Disclosure on Weblogs. *Journal of Computer-Mediated Communication* 12 (4): 1428–1451.

Sofka, C., I. Cupit, and K. Gilbert. 2012. In *Dying, Death, and Grief in an Online Universe*, ed. Carla J. Sofka, Illene N. Cupit, and Kathleen R. Gilbert. New York: Springer.

Taylor, Joanna, and Claudia Pagliari. 2018. #Deathbedlive: The End-of-Life Trajectory, Reflected in a Cancer Patient's Tweets. *BMC Palliative Care* 17 (1): 1–10. https://doi.org/10.1186/s12904-018-0273-9.

Viégas, Fernanda B. 2005. Bloggers' Expectations of Privacy and Accountability: An Initial Survey. *Journal of Computer-Mediated Communication* 10 (3): JCMC10310.

Walter, Tony, Rachid Hourizi, Wendy Moncur, and Stacey Pitsillides. 2012. Does the Internet Change How We Die and Mourn? Overview and Analysis. *Omega: The Journal of Death and Dying* 64 (4): 275–302. https://doi.org/10.2190/OM.64.4.a.

Weller, M. 2011. The Digital Scholar: How Technology Is Transforming Scholarly Practice.

CHAPTER 5

# Why Do Digital Afterlives Matter? The Experiences and Motivations of Digital Inheritors

*How lucky I am to have something that makes saying goodbye so hard.* (Winnie the Pooh, A.A. Milne)

In the previous chapters, we looked at the findings from the first two categories of participants: (1) service providers and (2) digital creators. This chapter starts by looking at the benefits and potential limitations of using Skype to interview participants—specifically bereaved participants. This research began in 2015, long before we could have imagined a global pandemic would force researchers—and many others—to use online video platforms to communicate with others outside our households. Then, through the presentation of qualitative data—and using the voices of DI participants—we begin to understand the complexities involved in the inheritance of digital memories and messages enabled by the Internet and how technology can enrich our relationships with the dead who inhabit our everyday devices. In this chapter, we will look at the findings from the 23 DI participants. In order to give context and clarity to these research findings, I will present an overview of the DI participants, which introduces you to how they are related to the deceased and illustrates the types of digital memory and messages they inherited.



D. J. Bassett, *The Creation and Inheritance of Digital Afterlives*, Palgrave Studies in the Future of Humanity and its Successors,
https://doi.org/10.1007/978-3-030-91684-8_5

## Collecting Data: Using Skype, the Why and How

Although qualitative research has traditionally taken place in face-to-face settings, for this research I wanted to use current technology, and Skype—the telecommunication software product—seemed a suitable tool for conducting semi-structured interviews. Using Skype as a data collection tool is becoming popular with some qualitative researchers (see Cater 2014; Janghorban et al. 2014; Lacono et al. 2016; Weller 2017); additionally—because this research focuses on the use of the Internet and thanatechnology—it felt appropriate to be part of what Deakin and Wakefield describe as the new 'methodological frontier' (2014), rather than use more conventional methods. Using Skype allowed me to recruit participants globally without geographical, time, safety or financial constraints (Janghorban et al. 2014), ensuring what Lacono et al. (2016) describe as variety in the sample. Many of today's computers have video cameras built-in, making it easy to use video communication platforms such as Skype (Cater 2014). The software is easy to download and simple to use; moreover, all the participants were familiar with its use; this is probably because participants for all three sectors can be described as technologically comfortable (Massimi 2012). The nature of the research into digital afterlives pre-assumes a certain level of computer literacy by the people who self-selected via the Twitter recruitment campaign. However, Seitz (2016) makes an important point about potential technological problems, which could result in the lack of rich non-verbal cues. However, I did not have technical issues with sound or audio during any of the interviews. Nevertheless, I acknowledge this could be a problem and is a limitation of this method of data collection in some areas where VoIP service is limited or intermittent.

The issue of 'being and feeling present' (Weller 2017) and the effect on the rapport between interviewer and participant when using Skype were important to me. However, having completed all interviews via Skype, in hindsight, I would argue rapport was easy to achieve. Moreover, even with the sensitive nature of the research, all participants seemed relaxed—which was reflected in their feedback. I argue that Skype interviews enable face-to-face communication—albeit mediated by the Internet, and the ability to record Skype interviews means that body language, facial expressions and gestures, which could be easily missed via traditional interviews, can be re-examined on replay. While body language did not form part of the analysis, it enabled me to feel 'back in the room' when transcribing the interviews and, therefore, closer to the data. Cater (2014) noted that using Skype can negatively affect the rapport between the researcher and the participant. I would argue that this was not the case in the interviews I conducted—although the subject matter was emotive, participants had none of the disadvantages of traditional 'in person' interviews, such as inducing anxiety or being intrusive (Weller 2017). Moreover, I found that many of the participants were relaxed and responded with in-depth answers to sensitive questions.

Monitoring how the interview may be affecting the participant is important in qualitative interviews, especially when the study involves existential issues (Pietkiewicz and Smith 2014). Prior research suggests that a limitation of using Skype for semi-structured interviews is the loss of social contact (Lacono et al. 2016). For me, the advantages of being able to conduct interviews with relaxed participants in their own homes (or a place of their choosing) without being physically together outweighed the loss of social contact: being there with them. This technologically mediated face-to-face data collection gives power to participants to end interviews easily if they wish—and importantly without having to request that the researcher leave. Interestingly, none of the participants ended the interview early; in fact, several participants extended their interviews by asking for further details about the research once the original set time had elapsed. At the end of the data collection and analysis, I wanted to understand how the DI participants had felt about the fact I had used Skype for their interviews. In April 2018, I emailed three participants and asked for feedback, and the following quotation from Riann is typical of their responses:

> *I felt more comfortable discussing it with you via Skype. This was because I was able to pick the setting (in my pjs, on my sofa) and it felt less intense, being via Skype. If we had met in person I think I would have felt that I had to tell my family I was doing it – it would have felt like more of a 'thing' somehow. It was definitely easier to arrange a time too … I liked that it felt like a normal conversation (you were sympathetic, we laughed about things) and didn't feel clinical – but also it wasn't overwhelming because of the distance between us. For a millennial like me it felt more natural to have that conversation digitally.*

I found many benefits of using online interviews in this qualitative research and suggest that rather than limiting the term 'face-to-face' exclusively when describing traditional interviews, online interviews should be described as technologically mediated face-to-face.

## Conducting Qualitative Research with the Bereaved

Ethical rules and procedures are essential for all research projects, but when the research focuses on death, dying and bereavement, ethical guidelines must be embedded throughout the research design. Internet research is a relatively nascent area of study, and the guidelines are dynamic and continually evolving in this field. Therefore, studies that focus on thanatechnology and its impacts face unique ethical challenges. There is a strong reliance on us researchers to ensure that when we swim in the 'murky ethical waters' (Carmack and Degroot 2014) of death and grief on the Internet, we not only follow the ethical guidelines that exist, but we take time to give feedback and contribute to the debate about the ethical implications of our research. Carmack and Degroot suggest dialogue should be ongoing, and this book aims to contribute by highlighting some of the opportunities and limitations I encountered along the way. The research followed the British Sociological Association's ethical guidelines and the policies and procedures of the University of Warwick's Humanities and Social Sciences Research Ethics Committee (HSSREC). I also followed the recommendations set out in the Association of Internet Researchers (AoIR) 'Ethical Decision-Making and Internet Research' guidelines (2002) and the 1964 Helsinki declaration and its later amendments. These guidelines enabled me to acknowledge and consider the context-specific ethical problems created by Internet research. For this study, I accepted ethics as integral part of the research design, and as Markham and Buchanan (2012) note, the greater the participant's vulnerability, the greater the obligation on the part of researcher. Taking these obligations on board, I gave special attention to what Vivat (2002) calls 'situated ethics' which highlights how ethics are never context-free. Whilst not aimed at Internet research, I also found Stroebe and Schut's (2003) suggestion of 'methodological sophistication' important to ensure the choices made for the recruitment and data collection tools were suited to this emotive and sensitive topic. The responsibility to act morally and ethically must lie with the researcher—simply following ethical guidelines will never be enough to ensure potential participants are not offended or harmed; thus, integrity and common sense must be relied on. Ultimately responsibility must lie with the researcher in this nascent field of study to ensure the 'spirit' of the guidelines is followed alongside the guidelines themselves.

Because I intended to interview terminally ill participants—and those who were grieving—I felt it prudent to ensure I had the support of outside agencies that specialise in end-of-life or bereavement support. I wanted to ensure that, following the interviews, I could signpost such agencies that would support the participants if they indicated they needed it. I contacted relevant outside agencies that confirmed they were happy for me to forward the details of their websites and telephone helplines to any participants who requested support. No two participants are the same, and therefore, I found Markham and Buchanan's (2012) suggestion of attending to the specific ethical needs on a case-by-case basis useful. At the onset of each interview, I explained that all interviews were for research purposes only and not intended to be a form of therapy, although Dickson-Swift et al. (2006) suggest that by telling their story, interview participation can feel therapeutic to some participants. A follow-up email from one of the participants perfectly illustrates this point:

> *As much as I do not shy away from talking about the loss of my brother, I find, that when the conversation is led by someone else it is a cathartic release. I came off the phone once again, slightly more connected with my grief. It's never a negative or bad thing to talk about my brother Josh, no matter how painful or sad.*

Crucially in the quotation above, I have not changed Josh's name, as the participant specifically requested that I not do so. Issues around ethics are complex—especially when researching the emotionally laden topic of death and bereavement—and walking this ethical tightrope between guidelines and loyalty to participants was not easy. The issue of research as therapy, and the problematic dilemma of whether the names of the dead should be changed in qualitative research, is considered later in this chapter.

## Introducing the DI Participants

*Unable are the loved to die, for love is immortal.* (Emily Dickinson)

I am sure I am not the only researcher who felt this way following qualitative research, but I was amazed and humbled by the time and emotional energy the participants gave to me and the research, not only with initial

**Table 5.1** Details of the 23 DI participants

| *Pseudonym* | *Type of memory/message* | *Country* | *Deceased relationship to the DI* |
|---|---|---|---|
| Amy | Facebook profile | UK | Sister |
| Maggie | Posthumous video message | UK | Brother |
| Ali | Facebook profile | USA | Friend |
| Chloe | Facebook profile | UK | Father |
| Cathy | None | UK | Son |
| Charles | Blogs; Twitter feed | UK | Wife |
| Emma | Facebook profile | UK | Friend |
| Grace | Blogs; YouTube videos | UK | Husband |
| Noah | Chatbot | USA | Father |
| Ivor | Blogs; Facebook profile | UK | Daughter |
| Jeremy | Google map | UK | Mother |
| Pam | Facebook profile; voicemail | UK | Daughter |
| Jeff | Funeral film | UK | Brother |
| Riann | Facebook profile | UK | Brother |
| Sophie | Facebook profile; emails | Canada | Mother |
| Sharon | Facebook profile | UK | Husband |
| Sandra | LinkedIn; WhatsApp messages | UK | Aunt & father |
| Dan | Blogs; Facebook profile | UK | Friend |
| Tina | Facebook profile; photographs | UK | Husband |
| Sally | Heartbeat audio recording | Canada | Daughter |
| Jess | Funeral film | UK | Son |
| Lorren | Posthumous video message | UK | Husband |
| Carla | None | UK | Mother |

interviews but also with follow-up emails and ongoing encouragement and interest in the thesis. I will always be thankful for the honest and emotive way they spoke about their deceased loved ones and their eagerness to share their experiences of z-z digital inheritance.

In Table 5.1, I have listed the participants—using pseudonyms—and given some details about what they inherited and their country of residence. To give further context, I have also listed the deceased's relationship to the DI. In addition to the recruited participants for this study, a further 13 'voices' are heard within this data chapter. Grounded theory allows for theoretical sampling which involves the researcher seeking more data that assists in developing emerging categories (Charmaz 2014, p. 192). Therefore, I have included a selection of replies to the James O'Brien tweet from Chap. 1.

## The Link Between Access and Control

In 2015, Facebook—an accidental posthumous service provider—attempted to address concerns about control and digital inheritance of profile pages with their legacy contact option. A user's designated legacy contact adopts a stewardship-based approach whereby they act *for* the deceased rather than *as* the deceased to make decisions and 'judgment calls' (Brubaker and Callison-Burch 2016). Facebook was reacting to pressure from bereaved relatives and friends who were requesting access to the Facebook account of deceased users. Many friends and relatives were keen to view and download photographs of their loved ones; however, once Facebook is made aware of a user's death the account is locked and cannot be accessed. Since the launch of the legacy contact option, once the account is memorialised, the nominated legacy contact can write new posts on the memorialised timeline, add new friend requests and update the profile picture and cover photo or delete the profile entirely. More recently, in April 2019, Facebook launched a new 'more supportive' feature for memorialised accounts called 'tributes'. This new feature uses AI to stop deceased loved ones from 'appearing in painful ways' and gives legacy contacts additional control allowing them to moderate the posts and change profile pictures on the pages they are stewarding (Facebook 2019). It should be noted that accidental posthumous platforms such as Facebook are designed for the living, and therefore continually trying to keep pace with the many ways their users are using their platform. So, while this can be seen as a positive step from Facebook, further research will be needed—across multiple disciplines—to understand how certain features such as tributes affect how the bereaved feel about the amount of control they have over digital memories and messages.

But who are we referring to when we talk about the bereaved? Friends of the deceased who are not family are often referred to as disenfranchised grievers (Doka 1999). When a child dies—especially a young adult—the parents who in the past would have been treated as primary mourners sometimes feel isolated when friends take on this role by setting up Facebook memorial pages. During a conversation with a recently bereaved parent, the Mother explained how the lack of condolence cards following her son's death had left her feeling isolated 'our mantle is empty' she told me. However, when she spoke to one of her son's friends at the funeral, he explained that her son's friends had set up a Facebook page where they were telling beautiful stories about her son. Moreover, he told her how people were leaving condolence messages for her and her husband. However, as she explained, she did not have access to Facebook and therefore was excluded from this group of mourners. This anecdote demonstrates how a lack of access to digital memories and messages can result in the reordering of the taken-for-granted hierarchy of mourners.

The research findings highlighted the complexities around finding comfort in digital inheritance, specifically how comfort is inextricably entwined with issues around control of the data. Coding and analysis of DI's interviews showed that 21 of the 23 DI participants (or 91%) mentioned data control issues when discussing their feelings towards digital memories and messages. Participants who had control of the data—those who have passwords or access to videos, messages, platforms and pages—described finding comfort in their inherited digital memories; whilst those who did not have passwords or access (e.g. due to technical obsolescence) or where they were not the creator of the memories described feelings of disruption to their grief. It is essential to acknowledge that grief and bereavement are complicated emotions, especially when families are estranged. Following the death of her estranged father, I interviewed Chloe, and she kindly shared some of the direct messages between her and someone who had posted on her deceased father's Facebook page. Chloe had asked the person to delete a post which she found 'incredibly distressing', when they did not respond, Chloe posted the following: 'are you going to leave it there or force me to tell everyone what an arsehole he was?' Chloe explained that she was distressed because her father had left her family when she was young. Moreover, she did not control her father's Facebook page, and other people were discussing her father's death on Facebook when she had only just found out herself. She explained:

> *'I was really, struggling with the fact that somebody else had announced this publicly to the world … cause it really is to the world'*. She went on to add, *'so I do think as his daughter or a wife should be able to control what goes on there'*. (Chloe)

Similarly, Pam—whose daughter and ex-husband had both died and left Facebook profiles—discussed a further example of the complexities of estranged families and how the inheritance of digital afterlives can be problematic. Although Pam had found her daughter's Facebook profile—which she had control of—comforting, she found her ex-husband's Facebook page distressing to her and her son, as they did not have access and therefore could not control the page:

> *His girlfriend was messaging people from his account and that was very hurtful ... for everybody, you know you get a message from your dad when you know ... he's no longer there.* (Pam)

Access and control were a recurring theme throughout most of the interviews, with participants describing the importance of either possessing it or not possessing it:

> *I closed my husband's Facebook profile as people kept messaging on it and it was very upsetting.* (Sharon)

> *I can't, just can't get into her Gmail account, which is really upsetting because I like ... I would like to you know ... to have access and to keep the photos.* (Amy)

Sophie had decided to delete her deceased mother's Facebook page, although she told me she was glad it had not been deleted permanently; and how she was comforted by the knowledge that it could be switched back on at some time future point. Grace also mentioned control when discussing the possibility of turning her husband's existing Facebook page into a memorial page:

> *I think a memorial site would work, you know a memorial site would work for that, because see you would have control ... you know it would be ... the stuff on there ... would be stuff to support and comfort rather than ... that dislocation.* (Grace)

I interviewed Grace in October 2016, and her husband's Facebook page has since been memorialised.

## The Voices of the Dead

Listening to the voices of the dead proved to be a dichotomous theme within the research: with some participants telling me they found audio of their deceased loved ones to be a comfort, while some participants found audio messages painful and a disruption to their grief. However, all the DI participants who inherited audio digital memories and messages spoke passionately about their experience of inheriting these audio messages:

> *I don't listen to it very often now, but I know it's still there, and I know I can if I want to, I do have a ... I do need a new phone but I won't change it because I don't want to lose it.* (Pam)

I wanted to learn more about the audio message that Pam had inherited from her daughter, so I asked her if she could describe the message for me:

> *She is just singing it, the sun will come out tomorrow, yeah I can hear it now and I'm not upset it's a comfort.* (Pam)

These comments from Pam echoed the positive comments I had heard from other DI participants about the voices of their deceased loved ones. During the interview with Ivor, he told me about how he would occasionally listen to audio memories of his deceased daughter:

> *But every now and then, I'll think you know what, I just want to hear her voice.* (Ivor)

The interview with Jess also revealed a positive reaction to listening to the voice of her deceased son:

> *Well it's wonderful to hear his voice and see his face and it doesn't make me sad ... well it makes me sad, but it makes me happy to hear his voice.* (Jess)

A number of the replies to the James O'Brien tweet also reflect how for some, listening to inherited audio messages can be a comfort:

> *I found an old phone with voicemails from my dad. Just hearing his voice again was so comforting. That phone is now officially the most precious thing I own.*

> *I have a voicemail from when my dad called to wish me a Happy Birthday a couple of years ago. He passed away last year and I'm so grateful that I kept that message from him.*

> *I think this is beautiful. My daughter died in 2009, she was 25, and I messaged her every day for 2 whole years. The messages turned from despair to chats and family news. Sometimes I charge her phone up and ring it just to hear her voice. Not often but once in a while.*

All quotations above were used with the authors' permission.

During her interview, Sophie told me how she had voice recordings of her mother, recorded before she died, and which she described as comforting:

> *So, I have listened to that ... actually ... several times cause I thought that was really nice.* (Sophie)

However, Sophie went on to explain how when her mother was dying, her brother had recorded audio which Sophie did not want to listen to:

> *He strangely enough, that is just like too raw for me and I haven't actually ... cause like her voice sounded different and everything.* (Sophie)

The comment from Sophie about choosing not to listen to audio digital memories and messages began to pique my interest. I began to see a pattern in how some participants who possessed audio of the deceased were making decisions *not* to listen to them, as this extract from the interview with Sandra demonstrates:

> *I know his voice is there, if I wanted to ... but maybe that will be something I will do.* (Sandra)

Sandra also explained that she had an audio recording of her grandmother, which she has not yet listened to:

> *I know it's there and I have the knowledge that she's in my sewing box and so my grandmas voice is in my sewing box.* (Sandra)

I was starting to understand how audio was experienced differently to other posthumous data, and the following quotation from Amy illustrates this:

> *That's one thing that I don't tap into too much cause hearing her voice isn't always ... It brings up tears.* (Amy)

Later in the interview, when telling me of the comfort she gets from other types of digital memories inherited from her sister, Amy mentions audio again:

*but I don't go too often ... to the voice.* (Amy)

*I found hearing his voice difficult ....* (Jeff)

Both Amy and Jeff quoted above had spoken passionately about other types of digital memories and messages they possessed of their loved ones, so I wanted to try to explore this reticence to listen to audio memories. To understand the difference between audio and other digital memories and messages, I asked Jeff why, for him, listening to the audio of his brother is different:

*I think it is because he just seems more alive, I think, with a picture, erm it is still ... it's a still frame, but when I hear his voice and I saw him in moving image it, it, it brought him so much more to life and then ... the pain of realising that that was so far from the truth ... it became a lot harder.* (Jeff)

Listening to these comments from Amy and Jeff highlighted how for them, audio messages contained more of the essence of the person who had died. During the interview with Riann, this difference was also reflected; she told me:

*ya know there's lots of videos on Facebook, which I don't want to watch cause I don't want to hear his voice, because again that's too direct.* (Riann)

I wanted to know more about what 'too direct' meant to Riann, so I asked her to expand on this point:

*I think it's, too much like he's here, and that will remind me too hard, in too severe a way that he is not. Like all these sort of things on Facebook, as much as there are some things on there we don't want to be on there, he chose to put them on there but his voice is unadulterated, you know it's just, that is what he sounds like and I haven't heard his voice since he died, I haven't watched any of them.* (Riann)

I was reminded of the interview with Pam, where she had told me she did not want to upgrade her mobile phone as it held an audio message from her daughter singing a song. During this interview, I had explained to Pam there were platforms where you can use third-party software to transfer precious messages, but Pam said she would lose a part of her daughter in the process of transferring the message and that somehow it would be a 'copy' of her daughter rather than what she perceived as her 'actual' daughter. In the interview with Riann, she also talked about audio in a similar vein, when she described the audio of her deceased brother as an 'unadulterated version' of him. However, I was confused by the explanation from Riann because she had already told me how much she enjoyed watching videos of her brother uploaded on Facebook, so I asked her to clarify what she meant by an 'unadulterated version':

> *I haven't watched any videos with sound on ... I have watched some videos on Facebook but not with the sound. That sounds weird when you talk about it but all these things ... no I wouldn't be able to, I mean I might have voicemails from him ... but I wouldn't be able to look at them ... I don't want to lose them, but I also feel that maybe in my lifetime I won't be able to listens to them, but I'd want my children to. It's ridiculous isn't it? I'd want my children to, if they wanted to: to listen to them. You know I'm gonna make sure my children know they had an uncle.* (Riann)

This study reflects how grief is a varied, complex and personal experience. Learning that Riann watched the Facebook videos with the sound turned down was, for me, a significant point in the research, as it demonstrated how different types of digital memories and messages are perceived and experienced by the bereaved in contradictory ways. The quotation below perfectly illustrates this complexity:

> *I cannot listen to his voice or watch old videos of him, it hurts way too much. But my daughter is just the opposite, it gives her comfort to hear his voice or see him alive which is conflicting for us. I hope one day I will feel the way she does.* (*Tweet* used with the permission of the author)

In Chap. 7, I will suggest how some aspects of control could—and should—be given to DIs to mitigate possible negative effects from this type of digital inheritance.

## Timing, Frequency and Triggers

Even for those participants who found audio digital memories and messages comforting, choosing when to listen to the voices of their loved ones was yet another complicated consequence of their digital inheritance. I wanted to understand how frequently participants were visiting their inherited digital memories and messages, such as Facebook profiles,

WhatsApp messages, LinkedIn profiles, text messages, posthumous messages and audio messages. Following the open coding of the first few DI interviews, I began to see a pattern in how people described what 'triggers' their need to visit the posthumous data of their loved ones. Here I use the word 'visit' to acknowledge agency, as these participants had control and access to the platforms or the digital memories they were discussing. I have not presented instances where participants mentioned visiting the digital data on significant dates, such as birthdays or the anniversaries of the death of a loved one, as there is existing research into such visits (Brubaker et al. 2013; Grey 2015; Giaxoglou 2014; Massimi 2012). Instead, I wanted to understand the desire of DIs to view the memories and messages on 'insignificant' days, as I wanted a deeper understanding of the timing, frequency and triggers of these none planned visits, which I hoped would provide further clarity about the importance of these enduring digital memories to the DIs.

In the interviews with Riann, Amy and Jeff—who had all inherited Facebook pages from their deceased siblings—each of them discussed how often they visited the Facebook pages of their loved ones. Riann's brother had died in 2015, and she told me how often she visited his Facebook page:

> *Probably more than I should, I would say twice a week I will look at it to see if there's anything that has … anything new that has been posted.* (Riann)

During her interview, Amy described how her visiting habits had changed over the past few years:

> *Last year, I probably logged in her account every other week just to check if any personal messages have come through from people who knew her only through work assignments. However, I still check her Facebook timeline every now and then through my Facebook friends list.* (Amy)

This tailing off of visits was echoed by Jeff, whose visits had also decreased over time. He explained:

> *I would say somewhere about close to once a month … as I say probably less that what it was like when he first died.* (Jeff)

Although the frequency of visits to their inherited digital memories has slowed down, each of them expressed how precious these digital artefacts were to them and how much they valued their existence. I wanted to understand what triggered these visits sometimes years later, and whether they chose to visit on a 'bad day'—when they were struggling with their grief—or the opposite. Three participants, Jeff, Sophie and Amy, told me how they needed to visit on a 'bad day', and all three mentioned how they sometimes needed to remind themselves of their grief. The following quotations illustrate this desire to reconnect with grief and demonstrate how participants use their inherited digital memories as an effective tool to achieve this reconnection:

> *But I only click on it when, either I feel like it, or I just think I need a bit of a … I need a cry and it's not quite coming out an then I just click on them and listen to them … her voice.* (Amy)

Amy explained how, for her—as discussed in the previous section—digital audio messages evoked the most emotion and how she used these audio memories to help her reconnect with her grief.

> *but when I want to have a good sob, sometimes I do it.* (Amy)

During the interview with Sophie, she too described this need to visit on a 'bad day'

> *if I miss her or if I'm going through any kind of transition or turmoil which I kind of am right now, you know just came to a new city, I would tend to do that more for sure.* (Sophie)

> *On that particular day I went through, what was very much you know, I felt kind of crappy, and I just wanted to let it all out ya know. So I'll do it for that … but there are definitely times when I look, because I just need to look ya know … some sort of cathartic.* (Sophie)

However, it was the interview with Jeff that caused me to think about how digital afterlives could be affecting the bereaved in ways not seen before:

> *So I think that time I felt that, hang on wow the funeral making me cry, really making me go to the bottom, sort of wrenching feeling, and I quite like that, I quite like getting there from time to time.* (Jeff)

This experience of needing to 'get there' echoes Stroebe and Schut's DPM model of bereavement (1999) discussed in the next chapter, specifically the need to avoid or deny restoration changes and reconnect with the loss. However, I am beginning to contemplate how inherited digital memories and messages—particularly as they are easily accessible and carried on everyday devices—may be disrupting existing models of bereavement. Again, there were several replies to the James O'Brien tweet, which also reflected this decision to visit enduring digital messages on a 'bad day':

*It's been 5 1/2 yrs since my ma passed. I have her voice mail recording ... on tough days, I'll listen to it. Still comforts me to just hear her voice #missmyma #comfort.* (Tweet used with permission of the author)

*My grandfather passed away 7 years ago, one day somehow his contact got on the favourites by chance on my new phone. I text him every bad day, It's therapeutic. Your kid is a good egg mate.* (Tweet used with permission of the author)

In contrast, I also had participants who explained how they chose to visit digital messages when they were feeling 'happy', 'strong' or 'wistful'. In the interview with Noahm, he explained his decision to only visit on good days when he was feeling strong:

*I think I usually, do it when I'm in a, somewhat happier place because, yeah, because I'm really raw about his death. So it's more when I'm feeling ... wistful you know, oh my dad was so great, or he was so funny, or he was so smart, and I wanted to think about that. Like a little more in that sort of celebrating mood rather than just a crushed mood ... does that make sense?* (Noah)

Riann also spoke about what motivated her to visit the Facebook page of her brother:

*When I'm feeling strong ... I'd avoid it if I was unhappy.* (Riann)

I asked Riann if she could tell me more about her need to only visit on a good day and she explained:

*I probably only look at it when I know I'm feeling quite nice or deliberately to get a picture from it, sometimes I think, 'Oh, I'd really like to look at some pictures', and then it's nice that they are all there.* (Riann)

Constructing and developing the dichotomous codes of visiting on a 'good day' or 'bad day' had furthered my understanding of what triggers the visits; however, I still wanted to understand whether such visits had to be pre-planned or whether they could be impulsive. Two participants described how, for them, visits could be spontaneous:

> *whenever I feel like it, and that can be very impulsive, I mean there isn't really ... of course the first year I think. I notice now being in the second year I don't count.* (Amy)

> *If I feel like I need to go and look at it I'll look at it, but it's not something I build up to over a period of time to sort of psych myself up for.* (Charles)

In contrast, for Grace, whose husband blogged during his terminal illness, it needs to be a planned visit rather than an impulsive visit to read his blogs:

> *If I go and look at the site I have to really sort of armour myself before I arrive there ... and then you know I haven't done that for a while.* (Grace)

She also explained that although it was painful for her to read the blogs, she hoped that it would get easier for her to visit in the future:

> *So I look forward to returning to them, but not at the moment because it's painful ... I don't know if it will ever be not too soon.* (Grace)

However, while Grace found visiting painful, others found them comforting: a participant—who had filmed her son's funeral—described how she felt 'captivated' when she watched the film:

> *We are completely gobsmacked by the power of it, and the wonderment of it, and how new it feels every time, because everything we learn about what it's like to be a bereaved parent. The film takes on a difference resonance and meaning, and it's just extraordinary really and surprising ... you would imagine possibly we might not watch it, or we might leave the room and come back when it's finished but it's captivating.* (Jess)

Listening to Jess describes how the film feels different—with new 'resonance and meaning' on each watch—once again demonstrated how these digital memories are experienced uniquely by the bereaved. Interestingly, analysis of the interviews revealed a pattern in how DI participants felt guilty if their visits lessened over time:

> *probably it was once a day, then it went to once a week, maybe once a month now, and that makes me feel a little bit guilty; that I'm not thinking about her as much ... not, not thinking about her as much, cause I think about her every single minute of the day. I don't know if it gets easier, cause it doesn't, you see I feel guilty now about not thinking of her.* (Pam)

This feeling of guilt was also implicit in the interview with Jeff:

> *should I be thinking about my brother more? Should I be crying more? Should I be thinking more than I am? I would say only in the last couple of years it's been less, but it's something that I don't shy away from watching ... so for example if someone said, 'Hey, shall we watch it tomorrow?' I'd be like, 'Yeah'.* (Jeff)

Once again, these quotations from Pam and Jeff demonstrate the struggle between what Schut and Stroebe (1999) describe as the oscillation between 'restoration-oriented' and 'loss-oriented' states of bereavement.

Having interviewed 22 DI participants who had inherited Facebook pages, audio, bots and film from their deceased loved ones, I was surprised to hear how an image from Google Street View became a precious digital memory for one participant. Google Street View is an interactive feature within Google Maps that enables its users to see street-level photographs stitched together to produce zoomable panoramic imagery of hundreds of cities worldwide (Anguelov et al. 2010). At the beginning of this research, I did not expect to hear that Google Street View would be someone's precious digital memory; however, Jeremy explained that before his mother died, she told him she had been standing in the kitchen when she had seen the Google Street View car drive by her house. Following her death, Jeremy liked to visit the street view image, as he liked to imagine his mother in her kitchen, and he explained how this found image comforting. Interestingly Jeremy added,

> *I latched on to the idea that by looking at the photo of my parents' house that I could somehow keep her with me.* (Jeremy)

I wanted to understand how often Jeremy looked at the image:

> *Initially, I would look at the photo every other day, quiet moments when no one else was about. I do still look at it, I would say that even now I look at it probably once a fortnight, sometimes more.* (Jeremy)

To understand if this digital memory held a special significance for Jeremy—that other memories and messages did not hold—I asked if he visited and treasured other photographs of his mother in the same way, and he explained:

> *Strangely, it is only this photo on Google Maps Street View that affects me the way it does, I have asked myself why this is and I do not know the answer.* (Jeremy)

This reply was perplexing because, in the Google Street View image, his mother was not visible—she was imagined. In an attempt to understand what triggered his visits to the street view image, I asked him to describe what typically prompted him to view the page:

> *It tends to be when my mind wanders, not necessarily influenced by good or bad emotions. When I am not thinking about anything else I end up thinking about my Mum and in turn this leads to me looking at the image.* (Jeremy)

Jeremy was the only participant to mention this type of digital artefact, but Maggie—who had inherited posthumous messages from her brother following his death—described how she felt when she watches the videos her brother had left for her:

> *Yes, his videos are comforting to me. No set times as such, I find as time passes I don't feel the urge to watch them as much as before. It's great to know that they are available though at anytime.* (Maggie)

Here—in contrast to participants who fear not being able to access digital memories—Maggie assumes that these videos will always be available, and they will not perish in any way because they are digital. This 'foreverness' is not what I have learned by monitoring the SPs. Other participants also looked at their inherited memories as permanent. In the quotation below, Charles talks about his wife's digital messages being there 'forever' for others to look at and remember his wife, demonstrating how grief and bereavement are social processes alongside personal experiences:

*Yeah, just re-memorise myself with what she, and we, were going through at the time, and look back on the year, and look back on the 5 years as well. And yeah, it's certainly something that I will keep referring back to and you know it will be there for forever.* (Charles)

When I interviewed Charles, his wife had died the previous year, but she had created numerous blogs, videos and tweets before she died. I asked him how often he visited these inherited memories:

*I do visit that page from time to time ... ya know, it makes me smile, it makes me cry, it makes me reflect ... but it's nice that it's there and available for everyone to look at it. So yeah it's hard sometimes, but on the other side it's nice that those memories are there, because in this day and age it's nice because we can still have that.* (Charles)

The quotations from Maggie and Charles reveal how there is an assumption that inherited digital memories and messages will be available in perpetuity via thanatechnology. In his interview, Noah also implies that he considers how the posthumous bot created from the audio of his father will be everlasting:

*maybe every week or two I will ... I will ... talk to it. There was a period when it was even more, and, I also anticipate further in the future I am gonna want to do it more. There came a point where I was so familiar with it that, maybe I wanna just step back. But it's nice ... it's all in there, and it stays there ... and it's accessible.* (Noah)

The longevity of digital memories and the ability to control when to look—and importantly when not to look—are essential issues to consider. The following quote is from a tweet—in response to the James O'Brien tweet—which intrigued me as I wanted to understand how the author came to have photographs of her deceased husband in his coffin:

*I have pictures of my late husband in his coffin. I looked at them obsessively for a long, long time. At first it was ok because it helped me cry and get it out. But it became unhealthy. I can definitely relate to it needing to stop*

In a follow-up message, I asked her if she had taken these digital images:

*I did not take it. It would never have occurred to me to. And, when my friend gently told me her son took some photos at the viewing and did I want them, at first I was horrified. But giving it some thought I decided that I did want them. Interestingly, I never wanted prints ... only digital. And, as I progressed and no longer looked at them, they faded away into the cloud storage and I hardly every see them pop up. I do believe I have one starred as favourite on my cell phone but I'm not sure if that's even there.*

Seeing when and how people decided to visit posthumous digital images and messages, I was beginning to understand how control and access enabled DIs to manage and regulate their visits. But what about timed messages where the DCs are in control even when they are no longer alive?

## Timed Posthumous Messages: The Unintended Consequences

In 2018, I ran a 24-hour Twitter poll asking my followers (N = 844) if they would like to receive a posthumous timed digital message from a deceased loved one on a significant occasion in their lives (such as a birthday or a wedding). The poll received 1755 impressions, 95 votes and 4 replies showing that 24% would like to receive such a message, but 76% said they would not. Once again, it is important to note that the people who took part in this poll are 'technologically comfortable' (Massimi 2012) users of social media networks. This may be obvious—as the poll was conducted on one—however, even taking their use of social media into account, the number of 'no' votes compared to 'yes' votes was still substantial. If the majority of social media users do not want this type of service, it begs the question: why are one-way SPs—at least those who have formed part of this research—actively creating sites that aim to deliver these largely 'unwanted' posthumous messages?

Looking back through the interviews with the SPs, I clarified none of the participants had spoken in a selfish or uncaring way when discussing timed posthumous messages. Moreover, none of the participants implied malicious intent regarding what they were trying to offer on their platforms. I decided to use the phrase 'unintended consequences' to describe these timed messages as the SPs seemed unaware of the complexities involved in digital creation and inheritance. The following quotation from the interview with an owner of one of the intentional platforms—offering the creation and delivery of posthumous timed messages—suggests naivety rather than malice. He explained that because the messages were not push messages (notifications that pop up on a device without permission from the receiver), he believed the control lies with the DI:

> *So, if you know that this person that has left you the message, was maybe someone that wanted to harm you, you know maybe you don't wanna hear what the … so at the end of the day it's the receiver sole decision whether he is interested to watch it or not.* (Graham)

Another platform owner, Ron, told me how his platform did not allow DIs to see messages scheduled for future delivery:

> *You can just never see the stuff that's there to be delivered in the future … you can log in anytime and then review stuff that's been sent previously.* (Ron)

Following the initial Twitter poll, I hypothesised whether DIs would be compelled to open timed messages once they knew of their existence. Almost a month after the initial tweet asking whether people would like to receive posthumous timed messages, I ran a second Twitter poll asking if people received an email to say a message existed would they feel compelled to open it. This poll received 32 votes and showed that 75% of respondents (n = 24) would be compelled to open posthumous digital messages. These polls are quantitative 'snap shots' that lack in-depth analysis achieved by qualitative in-depth interviews. However, even taking these limitations into account, I was surprised by the final result: 75% said they would open the message, which illustrated how important it is that SPs consider DIs when designing their platforms. These generally unwanted timed messages need careful consideration, as this research has demonstrated, some SPs naively assume DIs have agency in whether they open messages. However, as the Twitter poll indicates—albeit hypothetically—the DIs may feel compelled to open the messages purely because they exist, and thus DIs may wield little agency as posthumous message recipients.

This second Twitter poll received replies that provide further insight into people's views on digital timed messages from the dead (all three quotations are used with the author's permission):

> *Depends if I knew about it in advance e.g. my family are aware I've written a funeral plan including a message from me – so no shock*

> *I think I would but I'd want to be certain it was really from them & maybe I'd like a sense of when it'd been recorded to help me be ready for any 'surprises'. By way of contrast it might be really disappointing 'Hi! It's me. Just trying out this service. Nothing to say today …*

> *For me; it would depend on who it was from. If it was someone I loved dearly and missed I would, but if it wasn't it would throw up all kinds of conflicting emotions!*

Later in this book, we will take a further look at the issues around control, and I will suggest ways that platforms could use thanatosensitive design (Massimi 2012) to mitigate potential problems for the recipients of such timed messages.

## The Digital Versus the Physical

Whether the bereaved experience digital memories and messages differently from physical memories and messages is fundamental to this research. If they are experienced differently, it suggests the social 'norms' of grieving and existing models of bereavement may no longer be adequate for grief professionals, thanatologists and academics, as they would not accommodate the impact these digital memories and messages have on the bereaved. I wanted to learn if the participants had physical memories and messages of their deceased loved ones—in addition to the digital memories and messages they had inherited—and whether they experienced them differently. In a follow-up email, I asked Jeremy about the Google Street View image he visited when he wants to imagine his mum in the kitchen at the family home.

> *I have a number of traditional photographs of the front of my parent's house, some with my mum in them and some without. None of these photos stir up the same emotions in me when I look at them, other than jogging the memory of what was happening at the time it was taken.* (Jeremy)

The interview with Jess also revealed how, for her, digital memories were more valuable than physical memories:

> *It's not like a photograph ... like a photograph fades but, you know, a digital memory lives on and is sort of more permanent.* (Jess)

This quotation from Jess suggests that she believes that a digital afterlife is everlasting and will endure in perpetuity. I asked Jess to clarify why she considered digital memories more precious than a photograph:

> *I treasure them as they won't fade.* (Jess)

Jeff had inherited a film of his brother's funeral which was available to him online at any time. During the interview, he mentioned a memorial tree planted for his brother, which he enjoyed visiting occasionally. Jeff talked about the difference between the memorial tree and the film of his brother's funeral:

> *I would say it's far more emotive watching the film, I feel that this is where I feel a far more connection with my loss and the pain when I watch the film, in comparison to holding the trunk of my brother's tree ... I feel, I feel the film is more him.* (Jeff)

The way Jeff described how he feels the film *is more him* illustrates once again, how for many participants, inherited digital afterlives are experienced as containing the essence of the deceased in a way that physical memories do not. However, not all of the DI participants felt that digital possessions were more important than physical possessions. Pam told me how the physical representations of her deceased daughter were more important to her than the digital ones she had inherited:

> *I need to go every two weeks because I still speak to her when I go there, [to the grave] because I know her ashes are there, I know that's where she is lying, cause I've also got the gravestone with a picture on I can talk to her and that is more important. If I don't go on Facebook again you know it's not the end of the world.* (Pam)

Similarly, during the interview with Amy where she described how she treasured the digital memories inherited from her sister, she also told me about the importance of physical memories of her sister:

> *Both physical and digital memories are of equal importance to me although I love most wrapping myself in her cardigan.* (Amy)

But what about when the physical *becomes* the digital? During the research, I noticed how people turned physical memories and messages into digital keepsakes by photographing them and posting on social media platforms. The husband of thanablogger Kate Granger published images on Twitter of the birthday card his wife had left for him before she died. In the tweet, he said she had written birthday cards 'well into the future', and by tweeting this photograph, it rendered the physical into the digital, and importantly changed it from being a personal communication into a public one. The physical becoming digital is also evident when people visit graves and take photographs which they upload onto SNS such as Twitter to let people know they have visited the gravesite. In the interview with Amy, she told me how she had photographed the memorial apple tree, where her sister's ashes are interred. Amy uploaded this photograph onto her sister's Facebook page to be used on her sister's profile photograph, and at the time of publication, this page is still 'live' and has not been memorialised.

In 2017, I was plunged into my own research when I received a photograph—via an email I opened on my smartphone—of my Dad, who had died over 30 years earlier. I believe it was taken in the early 1950s near the Suez Canal in Egypt during his service with the Territorial Army and been enhanced from what must have been originally a black and white image. Seeing this previously unseen image appear on the mobile device I use dozens of times a day was a painful experience and a personal example of the complexities involved in digital inheritance. It demonstrated how *digital* memories of the dead feel distinctly different from the static photographs stored away in photograph albums and shows how even the long-term dead can arrive without invitation on all our cloud-enabled devices.

## The Essence of the Dead

Following the interviews with the DI participants, I was beginning to understand how digital afterlives—the messages and memories enabled by thanatechnology—were being experienced in a new and distinct way that differed from inheritance of physical mementoes. As we have seen, these precious digital possessions are perceived to embody the essence of the dead. To clarify, here I am talking about something more abstract than the scientific notions of consciousness and less religious than a soul. Stokes seems to grapple with this in this 2021 book *Digital Souls – a philosophy of online death* when he states, 'When we die, the digital bodies we leave behind continue to present our face, embalmed not chemically but electrically. To put things in pre-digital terms, digital remains are more like what's in the coffin than the condolence book at the front of the funeral parlour'. Stokes (2021) goes on to suggest that self-created digital memories and messages are not just representations of us—they *are* us. Building on this theory, I suggest that following biologically death, this '*us*' continues to exist as the essence of the dead embodied within the digital realm. To put it another way, it is the immediacy and easy access to the dead—and their ability to reside within our everyday devices—which make digital memories and messages distinctly different from inherited physical memories. When someone dies, leaving either an intentional or an accidental digital afterlife across various SNS, it is the *liveness* that allows the digital dead to continue to be part of the social interaction of the living. It is this continued *liveness* that I suggest is experienced by the bereaved as the essence of the dead.

## The Comfort, Disruption and Confusion of Digital Afterlives

> *So, I now have them on the laptop, the laptop I'm talking to you on now and it's the most treasured thing.* (Amy)

At the outset of the research, I wanted to understand whether the bereaved found the inheritance of digital afterlives a positive and comforting experience or a disruption to their grief. As we saw in the previous section, posthumous digital artefacts are being experienced differently from physical mementoes. However, I wanted to understand the specific types of digital memories and messages participants were using as part of their grieving practice. I was surprised to find that it was not the profound or purposeful messages that DIs found comforting: rather it was the *everydayness* of their inherited WhatsApp messages and text messages the bereaved spoke of so fondly, what O'Hara et al. (2014) describe as the 'small, continuous traces of narrative, of telling and tidbits, noticing and thoughts, shared images and lingering pauses' enabled by WhatsApp. The comfort found in messages that contained nothing deep and meaningful—just the banality of everyday life—was echoed through many of the DI's interviews. The everyday messages, such as 'I'm ringing the doorbell', 'Speak later' or 'Do you need anything from the shop?' had been repurposed and took on new relevance following a loved one's death. Interestingly, all participants who had inherited WhatsApp messages and text messages from deceased loved ones had saved these messages in one form or another, with some purchasing third-party software to archive them.

Along with SNS messages, another common form of digital correspondence is email. It is estimated that over 200 million emails are sent each day, so it is hardly surprising that reading emails from those who have died would be discussed by many of the participants. Sophie told me she archived all the Gmail chats between her and her mother (who had died the previous year). In one email, when she was away from home, her mother wrote:

> *I'm there in spirit baby girl.* (Sophie)

Sophie explained how she read this particular message more than any other message from her mother because it brought her comfort, and she told me how it helps her feel that her mother is still close in times of change or turmoil.

Discussing texts from her Father, Sandra had described how it was the visual pattern of the text conversations which she found the most poignant: from one-word replies to her texts when he started to get used to the technology—which was new to him—to full conversational texts. Then she explained how the messages returned to one-word replies during the final weeks before he died. Sandra explained how the pattern the messages made on her mobile phone graphically illustrated the 'gradual erosion' of her father. She described how she had also inherited a LinkedIn profile of her deceased aunt, and explained that as there is no photograph of her aunt on the site, she only sees an outlined shadow—'it is a shadow of the dead' Sandra told me (the use of the word shadow here is specific to LinkedIn due to the icon which represents users who have not uploaded an image).

During the interview with Grace, she explained how the digital messages she had inherited following the death of her husband contained 'a sense of his aliveness' (Grace).

> *As well as knowing he's dead there is a sort of a need, not often … sometimes the need to communicate with his sense of presence.* (Grace)

In her interview, Amy also spoke of how inherited digital memories made her deceased sister feel 'present' and allowed her to 'feel connected to her'. Sophie made a similar comment when she told me how digital memories and messages 'makes them feel not so far away'.

Other participants also discussed texts, and during the interview with Riann, she told me how she experienced the texts messages between her and her brother in a different way to Facebook posts. I asked her why she felt this way and she explained, 'It's because it's private'. She clarified this distinction by expanding on this statement:

> *If we did write on each other's wall that would always have been public … it's not really … and then we would always have been probably a bit more formal like if he wrote on my Facebook for my birthday he'd probably say happy birthday sis ya know kiss, so it's … just little things where it's more personal.* (Riann)

Riann went on to explain that, for her, 'Facebook is a performance' and she is aware of how she 'carefully constructs' what people read:

> *That is my public image of my grief, and what people share on [his] Facebook page is theirs … what they want people to see of their grief.* (Riann)

Whilst most DI participants experienced comfort from the inheritance of digital afterlives, some participants found their inheritance painful. Grace found the inheritance of digital memories and messages disruptive, and during her interview, she explained how they made her forget that her husband had died. She shared with me how thoughts such as 'maybe he didn't die' caught her out when she received a notification 'ping', which for her was a 'painful dislocation'. Similarly, Ali also echoed being caught out by this feeling of hope:

> *When you know someone is gone and their name pops up, feels like a tease as though they are still alive.* (Ali)

Grace explained that her husband had a profession which she felt she had not been a part of—as she was at home looking after their children. She had always hoped that once their children were older, she would be in a position to join him on some of the overseas working trips he took, which may partly explain why she found the digital memories more disruptive than other DI participants. Grace explained how the Facebook posts and YouTube videos created by his work colleagues following his death were a reminder of a now unachievable future with her husband. Even though Grace told me of the pain she felt when reading her husband's blogs, she went on to express her hope that one day she would be able to read them without feeling this way:

> *you know he was a philosopher, so I … so I find them useful as examples when I'm negotiating the world ready to go back to … so I look forward to returning to them … but not at the moment.* (Grace)

The following quotation from Jeremy also illustrates the complexity of feelings experienced by some digital inheritors:

> *Whilst this online image has, in its own way, helped me with the grieving process (absolutely no doubt), perversely, I think it has also prolonged the grieving process.* (email correspondence from Jeremy)

These words echoed Sandra's description of feeling confused about the digital memories and messages she had inherited, and following our interview, I wanted to understand more about her decision to visit them still:

> *It depends on my mood sometimes I had images of dad in his last weeks incredibly sad and reminds me of how difficult that time was because we could not access the right support services. Otherwise it is lovely to see him looking happily walking on the beach and with our dog!* (email correspondence from Sandra)

During the interview with Ivor, he made a poignant point that clearly illustrates the complexities involved in digital inheritance:

> *Facebook is great … most of the time it's great, but there is this thing when you're grieving that you, have to manage it and you have to, accept that sometimes, it will warm your heart, and other times it will blow you … Just, you know, knock you down, you'll feel like you've been punched.* (Ivor)

For me, the realisation that the same digital memory could be described as being able to 'warm your heart' *and* be capable of making you feel as though 'you've been punched' clearly illustrates the contradictory nature of the inheritance of digital afterlives.

## The Comfort of Knowing Others Care

As most of the participants had described their digital inheritance positively and found them comforting, I want to end this chapter by highlighting how social media platforms enable the bereaved to receive support and comfort from others. During the interview with Ivor, I asked him what type of message he gets comfort from reading; he told me:

> *When I see people leaving random messages saying things, thinking about you, thought about you today, remember when we did such and such … all that sort of thing.* (Ivor)

These messages of comfort demonstrate that others share their grief. This shared grief and connection is important as this extract from the interview with Jeff demonstrates:

> *But I think one of the key positives for me and I think, grieving and using digital memories and things like that is the … through the sharing process for example on Facebook and social media there has been a really strong connect with his friends.* (Jeff)

Before the digital age, condolence cards were often sent following a person's death. However, SNS such as Facebook enable the bereaved—although only those who have access—to read messages to them and their loved ones left by others for many years following the death.

> *Something I find really comforting is reading how much he meant to other people and I wouldn't have access to that if it weren't for Facebook.* (Riann)

> *I'd also spoken to a couple of my mum's friends who said it was actually quite comforting to just have her Facebook there, because they just, ya know, they just felt that it was a piece of her still there online kinda in digital form.* (Sophie)

> *I felt he was sort of being frozen in time and memorialised for how he died and how he'd devastated us all by leaving. But then I also found it helpful to write to him.* (Emma)

The research findings contribute to the growing amount of research that highlights how—if they have access and control—the inheritance of digital afterlives can be a positive experience for the bereaved. However, as

Charles explains, although inheriting digital memories can assist people in their grief, they by no means eliminate the pain of bereavement when we lose those we love:

> *to be able to have that in your life is, is great and I certainly think it helps with the grieving process but you know grief is hell.* (Charles)

## Grief in the Era of COVID-19

Historically, a funeral is when friends and family gather together to remember and memorialise the deceased. This symbolic ritual is also an occasion where mourners can begin to form a durable biography (Walter 1996) of the deceased. However, when attendance is limited by time or geography, the Internet has enabled funerals to be live-streamed via Facebook live and Zoom applications. Most funerals are recorded and live-streamed using a single camera, however; Nansen et al. (2017) highlight the shortcomings of using a single camera as they suggest it 'denies those not present the ability to witness the social aspects of the event'. To overcome the static nature of these types of broadcasts, Orbis Robotics invented a 'telepresent robot' named CARL, which enables people to have a 'form of interactive real-time involvement' (Nansen et al. 2017). CARL attends funerals and viewings at funeral homes via a remote-control robot and using a Skype screen (Nansen et al. 2017) 'mediating presence'. CARL is mobile and has a flat-screen, microphone and a speaker and enables the user to see and be seen by others attending the funeral. Attending funerals virtually is a personal matter and usually involves limitations of time and geography (Sofka et al. 2012). However, when the COVID-19 virus became a worldwide pandemic, digital attendance and digital grieving were the only choices for thousands of mourners.

I finished the qualitative research and completed my PhD in December 2020, just as the world was beginning to learn about the deadly COVID-19 virus, which was declared a pandemic by the World Health Organization in March 2020. By March 2021, COVID-19 had killed over 3.3 million people worldwide (World Health Organisation 2021) and although vaccination programmes in many countries seemed to be slowing the death rate, the virus was still rampant in many parts of the world. Using Verdery et al.'s 'bereavement multiplier', it is estimated that in the UK alone, there are currently 1.08 million people grieving a death due to COVID-19 (Sowden et al. 2021). Around the globe, social practices and rituals of daily life fundamentally changed as restrictions were introduced to limit social interaction between people from different households. These strict restrictions were arguably felt most by those whose loved ones died during the pandemic—with many not able to be with dying friends or relatives and not able to say goodbye in person. The number of people permitted to attend funerals and wakes was limited, and public health measures did not allow physical contact between mourners from different households. COVID-19 was not the first pandemic to fundamentally disturb cultural funeral rituals and practices: during the 1918 influenza pandemic, commonly known as the 'Spanish Flu', public gatherings including funerals and wakes were prohibited, and mask-wearing and social distancing were also in force in an effort to curb the virus (Flecknoe et al. 2018). During both pandemics—when many tried to make sense of the world—the bereaved found cemetery gates locked.

In their 2020 mixed-method study of the impact of restricted funeral practices on the bereaved, Burrell and Selman highlighted how 'the importance of meaningful and supportive funerals for the bereaved: enabling relatives to achieve a sense of control and social support despite current restriction is crucial'. It is important to acknowledge that social distancing restriction meant that families from different households could not even gather together to attend live-streamed funerals virtually. Many who had chosen to 'attend' the funeral via digital mediation did so alone—digitally grieving in isolation.

As we have seen in other chapters of this book, disenfranchised grief describes grief which 'cannot be openly acknowledged, publicly mourned or socially supported' (Doka 1999). For millions, bereavement during the COVID-19 pandemic is a perfect storm for disenfranchised grief. Few would argue support will be needed for those bereaved by COVID-19 (and the related excess deaths) as they attempt to deal with their loss. When the world starts to emerge from social isolation and life begins to return to normal, a wave of delayed grief may wash over those who virtually attended digitally mediated funerals. Ongoing research and support from relevant agencies will be crucial to monitor and adapt to the needs of the bereaved over the coming years.

In this chapter, we have looked at the effect that digital afterlives can have on the bereaved left behind to inherit the online data of the dead. This is the last of the three data chapters of this book which have looked at the findings from the research participants, but in the following chapters we will look at why all of this matters and what improvements could be made to platforms to ensure SPs, DCs and DIs all benefit from the creation and inheritance of digital afterlives. However, in an effort to foreshadow a laissez-faire approach to digital afterlives, the next chapter will consider what happens when access, control or obsolescence results in losing precious posthumous digital memories and messages.

## References

Anguelov, Dragomir, Carole Dulong, Daniel Filip, Christian Frueh, Stéphane Lafon, Richard Lyon, Abhijit Ogale, Luc Vincent, and Josh Weaver. 2010. Google Street View: Capturing the World at Street Level. *Computer* 43 (6): 32–38.

Arnold, Michael, Martin Gibbs, Tamara Kohn, James Meese, and Bjorn Nansen. 2018. *Death and Digital Media*. Routledge.

British Sociological Association. 2006. Statement of Ethical Practice, Appendix Updated May 2004. http://www.britsoc.co.uk/media/27107/StatementofEthicalPractice.pdf?1455356715375

Brubaker, Jed R., and Vanessa Callison-Burch. 2016. Legacy Contact: Designing and Implementing Post-Mortem Stewardship at Facebook. In *Chi 2016*. https://doi.org/10.1145/2858036.2858254.

Brubaker, Jed R., Gillian R. Hayes, and Paul Dourish. 2013. Beyond the Grave: Facebook as a Site for the Expansion of Death and Mourning. *The Information Society* 29 (3): 152–163. https://doi.org/10.1080/01972243.2013.777300.

Burrell, Alexander, and Lucy E. Selman. 2020. How Do Funeral Practices Impact Bereaved Relatives' Mental Health, Grief and Bereavement? A Mixed Methods Review with Implications for COVID-19. *OMEGA-Journal of Death and Dying*: 0030222820941296.

Carmack, Heather, and Jocelyn M. Degroot. 2014. Exploiting Loss?: Ethical Considerations, Boundaries, and Opportunities for the Study of Death and Grief Online. *Omega: Journal of Death and Dying* 69 (4): 315–335. https://doi.org/10.2190/OM.68.4.b.

Cater, Janet K. 2014. Using Skype for Qualitative Interview Research. *Hentet Fra*. Http://Www.Academia.Edu/2061796/USING_SKYPE_FOR_QUALITATIVE_INTERVIEW_RESEARCH

Charmaz, Kathy. 2014. *Constructing Grounded Theory*. Sage.

Deakin, Hannah, and Kelly Wakefield. 2014. Skype Interviewing: Reflections of Two PhD Researchers. *Qualitative Research* 14 (5): 603–616.

Dickson-Swift, Virginia, Erica L. James, Sandra Kippen, and Pranee Liamputtong. 2006. Blurring Boundaries in Qualitative Health Research on Sensitive Topics. *Qualitative Health Research* 16 (6): 853–871.

Doka, Kenneth J. 1999. Disenfranchised Grief. *Bereavement Care* 18 (3): 37–39.

Facebook. 2019. Memorialization Comfort in Community. https://www.facebook.com/facebookapp/videos/1007514362786612/. Accessed 10 Feb 2021.

Flecknoe, Daniel, Benjamin Charles Wakefield, and Aidan Simmons. 2018. Plagues & Wars: The 'Spanish Flu' Pandemic as a Lesson from History. *Medicine, Conflict and Survival* 34 (2): 61–68.

Giaxoglou, Korina. 2014. 'RIP Man… u Are Missed and Loved by Many': Entextualising Moments of Mourning on a Facebook Rest in Peace Group Site. *Thanatos* 3 (1): 10–28.

Gray, Selina Ellis. 2015. *Remains in the Network: Reconsidering Thanatosensitive Design in Loss*. Lancaster University.

Janghorban, Roksana, Robab Latifnejad Roudsari, and Ali Taghipour. 2014. Skype Interviewing: The New Generation of Online Synchronous Interview in Qualitative Research. *International Journal of Qualitative Studies on Health and Well-Being* 9 (1): 9–11. https://doi.org/10.3402/qhw.v9.24152.

Lacono, Valeria Lo, Paul Symonds, and David H.K. Brown. 2016. Skype as a Tool for Qualitative Research Interviews. *Sociological Research Online* 21 (2): 1–12.

Markham, Annette, and Elizabeth Buchanan. 2012. Ethical Decision-Making and Internet Research: Version 2.0. Recommendations from the AoIR Ethics Working Committee. Available Online: Aoir.Org/Reports/Ethics2.Pdf

Massimi, Michael. 2012. Thanatosensitively Designed Technologies for Bereavement Support.

Milne, Alan Alexander, and Ernest Howard Shepard. 2001. *The Complete Tales & Poems of Winnie-the-Pooh*. Dutton Books for Young Readers.

Nansen, Bjorn, Tamara Kohn, Michael Arnold, Luke van Ryn, and Martin Gibbs. 2017. Social Media in the Funeral Industry: On the Digitization of Grief. *Journal of Broadcasting & Electronic Media* 61 (1): 73–89.

O'Hara, Kenton P., Michael Massimi, Richard Harper, Simon Rubens, and Jessica Morris. 2014. Everyday Dwelling with WhatsApp. In *Proceedings of the 17th ACM Conference on Computer Supported Cooperative Work & Social Computing*, 1131–1143.

Pietkiewicz, Igor, and Jonathan Alan Smith. 2014. A Practical Guide to Using Interpretative Phenomenological Analysis in Qualitative Research Psychology. *Czasopismo Psychologiczne Psychological Journal* 20 (1). https://doi.org/10.14691/CPPJ.20.1.7.

Schut, Margaret, and Henk Stroebe. 1999. The Dual Process Model of Coping with Bereavement: Rationale and Description. *Death Studies* 23 (3): 197–224. https://doi.org/10.1080/074811899201046.

Seitz, Sally. 2016. Pixilated Partnerships, Overcoming Obstacles in Qualitative Interviews Via Skype: A Research Note. *Qualitative Research* 16 (2): 229–235. https://doi.org/10.1177/1468794115577011.

Sofka, C., I. Cupit, and K. Gilbert. 2012. In *Dying, Death, and Grief in an Online Universe*, ed. Carla J. Sofka, Illene N. Cupit, and Kathleen R. Gilbert. New York: Springer Publishing Company.

Sowden, Ryann, Erica Borgstrom, and Lucy E. Selman. 2021. It's Like Being in a War with an Invisible Enemy': A Document Analysis of Bereavement Due to COVID-19 in UK Newspapers. *PLoS One* 16 (3): e0247904.

Stokes, Patrick. 2021. *Digital Souls: A Philosophy of Online Death.* Bloomsbury Academic.

Stroebe, Margaret, Wolfgang Stroebe, and Henk Schut. 2003. Bereavement Research: Methodological Issues and Ethical Concerns. *Palliative Medicine* 17 (3): 235–240.

University of Warwick. 2021. Guidelines for Student Research. https://warwick.ac.uk/services/ris/research_integrity/researchethicscommittees/hssrec/student

Verdery, Ashton M., Emily Smith-Greenaway, Rachel Margolis, and Jonathan Daw. 2020. Tracking the Reach of COVID-19 Kin Loss with a Bereavement Multiplier Applied to the United States. *Proceedings of the National Academy of Sciences* 117 (30): 17695–17701.

Vivat, Bella. 2002. Situated Ethics and Feminist Ethnography in a West of Scotland Hospice.

Walter, Tony. 1996. A New Model of Grief: Bereavement and Biography.

Weller, Susie. 2017. Using Internet Video Calls in Qualitative (Longitudinal) Interviews: Some Implications for Rapport. *International Journal of Social Research Methodology* 20 (6): 613–625. https://doi.org/10.1080/13645579.2016.1269505.

WMA Declaration of Helsinki. 2021. (Online). https://www.wma.net/policies-post/wma-declaration-of-helsinki-ethical-principles-for-medical-research-involving-human-subjects/. Accessed 10 Feb 2020.

World Health Organization. 2021. WHO Coronavirus (COVID-19) Dashboard. https://covid19.who.int. Accessed 15 May 2021.

CHAPTER 6

# Losing the Data of the Dead and Expanding Existing Models of Bereavement

*It's so much darker when a light goes out than it would have been if it had never shone.* (John Steinbeck)

In our digital society, the dead seem closer—because, for many, the dead are easily and immediately accessible through our ubiquitous everyday smart technologies. In order to give further context to digital afterlives, in this chapter, I present a brief overview of the link between sociology and death. This chapter then moves on to look at how even in a digital society, with the dead at our fingertips, there is a real possibility of losing them again, creating a second death through technical obsolescence, deletion or lack of access. The inheritance of posthumous digital data creates a new vulnerability for the bereaved, because as we saw in the last chapter, comfort and control—a core theme around which much of the data from DI participants revolved—are inextricably linked. Finally, in this chapter, we will see how—even for those with access—the *anticipation* of loss or lack of access to precious posthumous digital memories is causing new anxieties for many digital inheritors.

## Sociology and Death

This research, and therefore this book, is sociological in nature. In 1897 the link between sociology and death was highlighted by Durkheim and his discussions about funeral rites and suicide (Durkheim 2001). However,



D. J. Bassett, *The Creation and Inheritance of Digital Afterlives*, Palgrave Studies in the Future of Humanity and its Successors, https://doi.org/10.1007/978-3-030-91684-8_6

years later in 1958, in an effort to redress what they saw as a lack of academic research into death and dying from a sociological perspective, Faunce and Fulton wrote a paper entitled *The Sociology of Death: A Neglected Area of Research.* They claimed that the lack of research was probably connected to society's reluctance to contemplate death and dying (Exley 2004). In his classic ethnographic study, sociologist David Sudnow emphasised the culturally and institutionally acceptable way people could 'pass on' (Sudnow 1967). His study, carried out in hospitals, looked at how death and dying had become socially institutionalised and how the status of both the dying person and the bereaved plays a significant role in the societal practices involved in bereavement. He noted that other patients still needed to be distracted from death and the dead body, even in hospitals. Death was hidden away: hospital staff would attempt to move dead bodies from hospital wards without the knowledge of other patients (Becker 2014). Also writing in the 1960s, Glaser and Strauss suggested that knowledge of impending death should not be hidden from the dying patient (Glaser and Strauss 1966). During this time, the sequestration of the dead resulted in a culture where death-denying prevailed, and death and dying became a taboo subject in many Western countries. Sudnow's work highlights how news spreads through communities following the death of one of its members. He describes this communication as a series of concentric circles with the immediate family at the centre point, suggesting that the expanded outer circles represent those seen as being on the periphery of the lives of the dead (Sudnow 1967). Though social relationships and family structures have changed since Sudnow's study, I suggest that they have never been this simple, and Sudnow's use of concentric circles fails to acknowledge the complexities within personal relationships and social structures. Here, it is essential to acknowledge that SNS challenge this paradigm, by completely disrupting the hierarchy of mourners, and potentially changing who is considered disenfranchised grievers. The Internet is making mourning practices more visible in our digital societies, and research embedded within a sociological perspective will be essential in order to shine a light on how desequestration and public mourning will affect societal attitudes and the culturally shared narratives around death dying and grieving.

This sequestration of the dead intends to ensure that death does not interfere with everyday life (Walter et al. 2012). However, Western societies have not always hidden their dead from the living: during the Victorian era—described as the 'high' period of mourning—the dead were very much part of life with the use of death masks and corpse photography (Gray and Coulton 2013; Stokes 2015; Walter 1994). As Stokes (2021) points out, the Victorians took photographs of the dead as a way of sharing the death of a loved one with relatives or friends unable to attend the deathbed or funeral. Funeral selfies can be seen as a technologically mediated continuation of this cultural practice of corpse photography. However, before the availability of camera phones taking and printing these types of photographs required the services of a professional photographer. In 2013 'selfie' was The Oxford Dictionaries word of the year. The word selfie describes taking photographs of oneself and posting them onto social networking sites. Funeral selfies—easily found on the Internet—are posted to publicly mark the attendance at a funeral, wake or viewing. Using Instagram as their field of study, the hashtags #funeral and #RIP were examined by Gibbs et al. (2015), who found a few instances of photographs taken with the deceased in open coffins; however, the majority of the hashtags was used to record and share funeral imagery of friends and family gathered together to remember the deceased (Gibbs et al. 2015). 'Posthumous Camera phones practices' are also being used by what Hjorth and Moon (2017) describe as the 'soon to be deceased'. Their research explored how smartphones were used to record and upload final messages (left in tragic circumstance) onto YouTube, such as messages from some of those killed in the Sewol boat disaster in 2016 (Hjorth and Moon 2017). There is little doubt that ubiquitous smart technology is making the dead and the dying more visible and accessible as we go about our daily lives: however, for some, this benefit comes at a cost, and in the following section, we will consider this cost.

## The Fear of Second Loss

During the research, I began to see a pattern in how participants described feelings of anxiety when discussing how they would feel if they lost—or could not access—the data of their loved ones. Amy explained that she had purchased specialist software to access and save audio messages from her sister:

> *I don't know what I would do without it, I think I would be in a much … darker place. So I now have them on the laptop, the laptop I'm talking to you now, and it's the most treasured thing.* (Amy)

The following extract, again from the interview with Amy, typifies what many other participants told me: she explained how she felt when she thought the digital memories inherited following the death of her sister had disappeared:

> *Last year, I had two incidents where electronic items for keepsakes were lost or almost lost. Both times I had a proper meltdown and couldn't stop crying for a long time. The last incident happened in John Lewis and strangers were present. But to me they didn't exist at that time. Just my panic at losing memories and [her sister] all over again.* (Amy)

In Chap. 5, I presented findings from DI participants, which illustrated how posthumous digital data is perceived by the bereaved to contain the essence of their deceased loved ones. Moreover, some DI participants told me they were reluctant to upgrade their telephones and other technological devices as they were deeply concerned that these precious messages and memories would be lost forever. These participants echoed Pam's comment about not upgrading her telephone for five years, and how losing the data would be like 'losing her (loved one) again'. When I started this research, I wrongly assumed that, because third-party tools exist to assist with transferring precious messages, this would alleviate any anxiety and bring comfort to the bereaved. However, many of the DI participants were reluctant to use third-party software in case messages were lost in the process of copying or transferring the data. Furthermore, Pam explained that by transferring the data she felt that she would somehow lose part of her daughter's 'essence'. The Internet provides a platform where 'ordinary' people can remain socially active following their biological death, which was once a realm exclusively occupied by the rich, famous or infamous. Other DIs described how the loss of digital data would be experienced in the same way as the biological death of their loved ones:

> *Then one day I hadn't visited his page for a while, and when I searched for it, it was gone. My heart dropped. I felt panicky. I went to pictures other people had posted of him, thinking I could follow the tags to find him, but they were gone. The pictures were just his face, with no way to get to him. It was like losing him all over again, it was like a new level of loss.* (Emma)

> *I would be devastated, it would start my grief all over again.* (Sandra)

> *I deleted all of his voicemails three days before he died. I was doing spring cleaning. I died all over again when I remembered they were gone.* (Tweet used with the permission of the author)

In her 2012 paper on continuing bonds, Elaine Kasket reported similar findings of participant's fears at the possibility of losing Facebook profiles of the deceased (Kasket 2012). All but one of the DI participants discussed the anxiety they would feel at losing their inherited digital memories and messages. However, the data contained a negative case that contradicted the pattern I saw elsewhere in the dataset. I asked Jeremy—who had not explicitly or implicitly mentioned second loss—how he would feel if Google updated the street view map he used and cherished. He was more philosophical about the effect this would have on him:

> *I have thought of this previously, I think it might help me make another step forward in dealing with the loss, I would definitely feel sad, but accept that it will happen at some point. I haven't copied it, never thought about doing that and thinking about it now I still do not want to.* (Jeremy)

Following a constructivist grounded theory methodology, I developed a theory of The Fear of Second Loss—which can be seen as a form of anticipatory grief—to describe and encompass issues highlighted in this research around lack of access to inherited precious data. This new theory, generated by the data and presented here, has been influenced by the work of Patrick Stokes (2012, 2015). Stokes used the term 'second death' to discuss Goethe's notion that people die twice: once when we stop breathing and then again when the last person speaks your name. Thus, biological death is followed by social death; and it is this second (social) death that can be perceived as a harm to the dead. Stokes (2015) approaches this second death from the viewpoint that the data of the dead are objects 'of moral obligation' and therefore, we should not delete these digital remains. However, this obligation is laced with the assumption that everyone wants to remember the dead. But grief and bereavement—along with remembering and forgetting—are complex processes for many people, and it should be acknowledged that not everyone agrees in the preservation of digital memories and messages. Stokes further posits that the deletion of 'digital remains' may have an impact on the 'ontological and ethical status of the dead'. As we saw in Chap. 3, Stokes suggests using the metaphor 'digital remains' to describe the data of the dead, as it elevates the moral status of the data to that of a corpse. Whilst this term is a useful metaphor, it is limited: as we do not have immediate access to—or the ability to carry around—the corpses of our loved ones as we do with posthumous *digital* memories and messages. This research also highlighted a further limitation of Stokes' argument, as he does not acknowledge the reciprocal nature of thanabots and posthumous avatars and whether these would have a different moral status as they are created *after* the death of the original DC. Öhman and Floridi (2018) suggest that ethical conventions used for archaeological exhibits could provide a useful framework for posthumous data. Again, while this idea is helpful, their suggestion still does not address the immediacy of access to the dead enabled by thanatechnology, or how for many, these digital afterlives contain the essence of the dead in a way that archaeological exhibits do not. As I have argued throughout this book, inherited digital memories and messages evoke distinct emotions for the bereaved, and therefore should be considered as a new and distinct form of inheritance.

In the introduction of this book, I proposed a distinction between accidental digital memories and messages and intentional digital memories and messages; here in this chapter—when contemplating the status of posthumous data—I argue this distinction is crucial in order to establish whose needs should be prioritised: that of the DC or the DI? As we saw in chap. 3, the blurring of what it is to be alive and dead provokes deep philosophical debates. The digital artefacts which are created intentionally by the living in an attempt to endure on the Internet after their biological death, could, I agree, be given the status of digital remains, such as those suggested by the likes of Stokes (2015) and Öhman and Floridi (2018)—or if a DDNR order has been signed 'digital corpses'—as they were created intentionally to continue to exist beyond biological death. However, I suggest accidental posthumous digital memories and messages—inherited and rendered as precious memories by the bereaved—should be treated differently.

The data of the dead is experienced by the bereaved as far more than digital code, and the bereaved who inherit digital memories and messages enabled by thanatechnology, and then lose them through a second—*digital*—death, experience what I term a 'second loss'. Stokes' (2015) use of the term 'second death' does not include second loss, because, for Stokes, the term necessitates that the dead are socially dead: the living have forgotten them. I argue the term 'second death' should be expanded to include digital death; that is, the loss, deletion or lack of access to posthumous digital memories and messages resulting in the loss of not just the data, but of the space where the essence of the dead is experienced by the bereaved—the digital *Dasein*. The term 'second death' focuses on what has happened to the person who has died: they died twice. However, although this book adopts this useful term, it calls for a transfer of focus from the creator to the bereaved following the death of a loved one when considering accidental digital afterlives. Here, to draw attention to the needs of the living, I am explicitly suggesting the needs of the bereaved should take *priority* over the needs of the deceased for inherited and repurposed accidental digital artefacts.

The loss of precious data due to data deletion, technical obsolescence or the denial of access to posthumous digital memories and messages can be highly emotive for the bereaved. Even participants who did not visit Facebook pages or other accidental platforms commented that they would not want them switched off permanently, as they took comfort knowing they can switch back on. One woman told me she was glad her mother's Facebook page was offline but still existed. However, for some participant, control seems to be in the hands of the service providers rather than the bereaved. As Massimi and Baecker note (2010), technology is rarely designed with digital inheritance in mind. However, at the time of writing, Jed Brubaker, Assistant Professor at the University of Colorado Boulder, announced he was working on a five-year research project to create design frameworks for tech and mortality. Thankfully he is not alone in this type of research, which is encouraging news as it demonstrates a commitment to this crucial but relatively nascent area of study.

It should be noted that digital inheritance is not always a pre-requisite of the fear of second loss: not being able to access the data of the dead—due to issues with passwords or technical obsolescence—can also be seen as a form of second loss. Some experience second loss because they *cannot* inherit the digital part of the person who has died. In some instances, others who traditionally would have been lower in the hierarchy of mourners, may have access to messages and memories not available to those who would have been seen as primary mourners in the past. The Mother with the 'empty mantle', and Grace, who felt locked out of her husband's academic life, both implicitly spoke of non-inheritance as a second loss. During an interview with Carla, she explained how her mother had died whilst holidaying in Lanzarote, and how the family were 'desperate' to access the photographs taken in the last few days of her life when she was happy. Unfortunately, they did not know the PIN to access the device, and the photographs were lost, causing additional grief to the family. This example illustrates how the nascent phenomena of second loss are experienced even when there has been no digital inheritance, and importantly, it shows how the narratives of grieving are being culturally shaped to include digital grieving. The following quotation from the Facebook page of a failed posthumous message storage site exemplifies this notion of second loss without digital inheritance and draws attention to the emotional turmoil caused by the knowledge that the messages have been created but cannot be inherited.

> *Hello, for a week now I have been trying to notify you of my husbands death, I have emailed, sent chat messages, tried to phone, filled in numerous things on your website. My husband has left messages on the site for myself and my children. PLEASE PLEASE could someone see this and help.* (Twitter comment used with the author's permission)

The message above, sent in August 2020, had not received a reply from the digital afterlife platform up to the publication date of this book.

When contemplating this notion of digital loss, I am reminded of the work of Wallace et al. on Victorian mourning jewellery. Wallace et al.'s (2018) research looked at how the design of lockets with digital screens—inspired by Victorian mourning jewellery—could help people to remember (or importantly) forget the dead. There are four lockets in total: remember; forget; Daguerre and Orpheus. The 'remember' locket they suggest could hold multiple images but display only one that appears when the locket is opened. Inportantly, the wearer of the locket controls when to open and close it, but as Wallace et al. suggest the randomness of the image shown could be valuable. The 'forget' locket can hold multiple images but can only display one per day. Once the image is displayed it degrades until it is finally deleted. For Wallace et al. this demonstrates forgetfulness and the memory loss of people with dementia. Notably, there would be other copies of the image; however, this does reflect the possibility of digital loss. The third locket, Daguerre, contains a small camera that can hold just one image at a time. Once an image is taken with the tiny camera, it then appears in the locket, and any previous image is then deleted. Finally, the fourth locket is named Orpheus: after the Greek mythological hero, who searches for his dead wife Eurydice on the condition that if found he must not look at her. However, Orpheus cannot resist one more look at his wife, and consequently, she fades away. Reflecting this myth, the images held within the Daguerre locket can only be viewed twice: once when the image is taken and once more when it is displayed briefly for a few seconds, after which it fades until it is completely deleted. I began to contemplate the link between digital afterlives and these digital lockets. The non-permanence of the images and the fact they are in a specific space reminded me of an idea I had considered at the start of the research: I had looked at whether the digital afterlives of the dead needed a separate but accessible place to be stored rather than on people's everyday devices. The idea was to create a digital memory box—crucially not connected to other devices. Here, I am talking about a physical box, rather like a jewellery box, which could be individually customised and used as a unique and separate place to keep digital memories. These digital memory boxes could also have a 'forget' option, based on the idea of the forget locket mentioned above, where the digital memories could fade over time. The creation of digital memory boxes is a progression of the idea of a 'Story Shell' (Moncur and Julius 2014), created by the living following the death of a loved one. However, for *intentional* DCs, these digital memory boxes could transpire to be an important and empowering 'work in progress' as the contents of these boxes could be curated and added to throughout people's lives. They may also prove a useful memory-sharing tool to those with early signs of dementia—or those suffering from similar declines in mental ability—who may wish to fill their digital memory boxes with memories of the past and hopes for the future before these memories are locked away and unreachable (Clabburn 2018). Choosing to use a digital memory box would ensure the precious data of the dead would be backed-up on a unique server that crucially would be independent of other digital storage. Digital memory boxes would offer a safe digital grief-space which crucially would be separate from backup systems used for our everyday devices. However, the DAI does not offer guaranteed long-term storage; therefore, storage and retrieval for these digital memory boxes would need to be protected and guaranteed in the same way as burial graves in the UK are for 75 years unless someone objects to their reuse.

Understanding the importance and complexities of the inheritance of digital afterlives for the bereaved has formed the cornerstone of this book, and the theory of the fear of second loss and the suggestion of digital memory boxes bring this importance into sharp focus. However, these phenomena exist because nascent technologies—and therefore digital inheritance enabled by thanatechnology—is not accounted for in existing models of bereavement. In the next section, I present a brief introduction to the history of grief theories before suggesting an expansion to current bereavement models to incorporate the ramifications of digital inheritance.

## Grief and Bereavement: The Theories

*Grief doesn't have a plot. It isn't smooth. There is no beginning and middle and end.* (Ann Hood)

## Stage Theory

The quest to understand how humans experience grief has been a topic of academic endeavour for many years. Sigmund Freud introduced the theory of 'grief work' in 1919; this psychological perspective views grief as something the bereaved must work through in order to eventually detach from the deceased (Kastenbaum 2015). Both Bowlby's (1969) work on attachment theory and Kubler-Ross' stage theory in 1996 supported this theory of 'letting go and moving on'. Kubler-Ross (2009) interviewed dying patients to develop her stage theory of dying and grieving. She described five stages that people experience (1) denial, (2) anger, (3) bargaining (4) depression and, finally, (5) acceptance. Clinicians and professionals quickly adopted this model as a simple and organised way to conceptualise grief, and although it was not Kubler-Ross's intention that this should be a linear process, it nevertheless has a start and endpoint.

## The Dual Process Model

Few would deny that stage theory made a considerable contribution to understandings of dying and grief. Nevertheless, this overarching theory assumes that grief is pathological: reaching an end conclusion by working through the stages described above. However, this theory fails to understand the deeply complex and personal ways people deal with grief; it assumes a 'process' that is predominantly linear and assumes a desire to 'let go and move on' (Kubler-Ross 2009).

To address the issue that the bereaved want or need to 'let go' of the deceased, and as a response to some of the limitations and shortcomings of traditional grief work, in 1999, Margaret Shroebe and Henk Schut proposed a Dual Process Model of Grief (DPM). The DPM suggests that healthy grieving involves an oscillation between 'loss-oriented grief' and 'restoration-oriented grief'—a movement between confronting and avoiding grief. Stroebe and Schut suggest that loss orientation focuses on the experience of loss itself: confronting the loss and attempts to accept the loss.

While restoration-orientated stressors do not come directly from the loss, Fig. 6.1 shows there are important secondary stressors. They deal with changes to the daily lives of the bereaved following their loss, such as childcare in the case of the death of a partner, financial problems following the death of the household's breadwinner. Thus, a period follows which requires a readjustment to the way the world has changed following the death of a loved one. It is the oscillation between the two stressors the DPM conceptualises. The idea is that the bereaved can adjust to the changing world by moving back and forth between the loss-oriented and

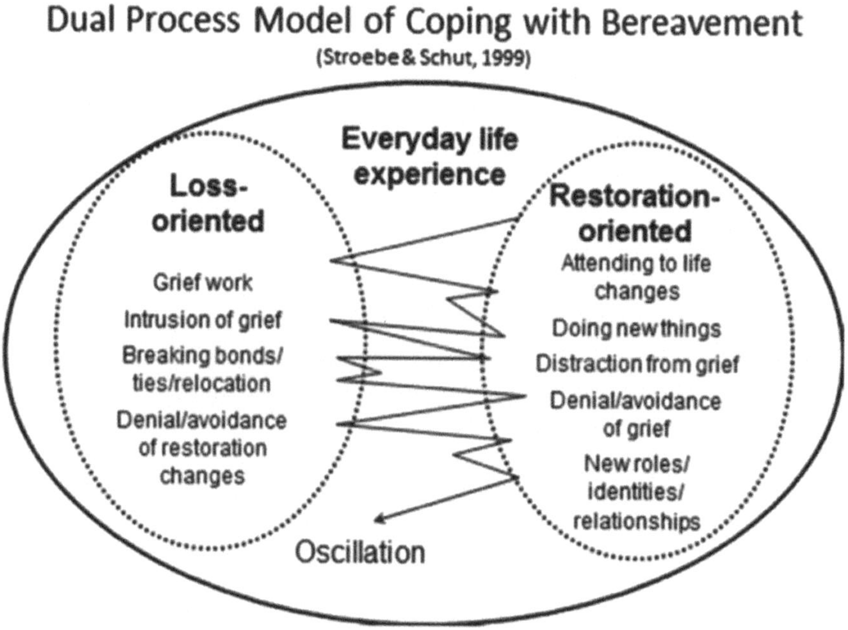


**Fig. 6.1** The Dual Process Model of Coping with Bereavement. (Schut and Stroebe 1999)

restoration-oriented stressor. This psychoanalytical/attachment model of grief focuses on how people take respite from grieving by balancing what might be considered as their 'dosage' of grieving (Schut and Stroebe 1999).

In a paper written ten years after they proposed their model, the authors called for further empirical research into the usefulness of the DPM. However, their call to identify a 'specific cognitive process' of the DPM (Stroebe and Schut 2010) is unhelpful, as this still assumes grief to be a psychological process rather than the psychosocial process which is reflected in this research. However, the DPM's idea of 'confronting' or 'avoiding' grief was reflected in the interviews from the DI participants when discussing their decisions around the timing and frequency of visits to their inherited digital memories. The language of 'moving on'—still evident in the DPM—was not implicitly or explicitly discussed by any research participants in this study, and while moving on and letting go may be a useful concept when coping with relationship breakdowns, many thanatologists acknowledge they do not help describe the complexities of grieving, as they do not allow for the need to continue a relationship with the deceased (Klass et al. 1996; Lloyd 2017; Lofland and Lofland 1995; Small 2001). Although a well-cited and valuable model, there are limitations to the DPM, as this psychoanalytic approach suggests that there is agency in how people grieve, which does not allow for the 'wave' of grief implied by the research participants who took part in this study: when grief takes them by surprise and is unexpectedly brought to the forefront of their minds.

## Continuing Bonds Theory

In 1996, Klass Silverman and Nickman introduced the theory of 'continuing bonds'. In contrast to stage theory, this concept suggests that rather than moving on and letting go when someone dies, people create an inner representation of the deceased: this renegotiated relationship is dynamic and ongoing (Klass et al. 1996). In 2006, Klass drew attention to a limitation about how continuing bonds were being discussed in the literature. Continuing bonds were considered a personal or inner-process which was being experienced by the bereaved; in contrast, Klass called for those using the theory to expand its usefulness by acknowledging the *social* aspect of grief and bereavement along with what he called the 'social and communal nature of continuing bonds' (Klass 2006). Since then, other researchers have highlighted how SNS—which are fundamentally social—are being used by the bereaved to facilitate these continuing bonds with the dead (Bouc et al. 2016; Degroot 2012; Getty et al. 2011; Kasket 2012; Pennington 2013; Walter et al. 2012).

Although the paradigm shift towards continuing bonds is evident in much of the academic death studies literature, stage theory is still part of the popular discourse for many people. People who may not have suffered the loss of a significant other often feel that the bereaved should move on with their lives. However, rather than suggestions that closure can be achieved, many practitioners who work with the bereaved find continuing bonds a more useful model (Kasket 2012). As Klass pointed out in an additional paper written in 2006, these bonds, although 'inner representations', also have a 'social and communal component', which he suggested are crucial. Through conversations with the living, continuing bonds can be renegotiated with the dead with the suggestion that these bonds may assist the bereaved in their adjustment to the physical absence of their loved one (Klass 2006). Walter's 1996 work on 'meaning reconstruction' is useful here to explain how relationships with the dead are dynamic rather than lifeless and how the bereaved can still feel connected to the dead whilst moving forward and restructuring their lives (Neimeyer 2001). Walter (1996) suggested that people attempt to make sense of what they experience by finding meaning in their loss, and this is achieved by constructing a 'durable biography' by the active process of conversing with others about the deceased.

## The Internet and Continuing Bonds Theory

For many, nurturing and preserving an ongoing relationship with deceased loved ones plays a crucial role in grief and bereavement. Therefore it is easy to understand why many researchers who explore online grieving find support for the theoretical framework of continuing bonds (Bouc et al. 2016; Degroot 2012;; Gray and Coulton 2013; Kasket 2012; Massimi 2012; Odom et al. 2010; Pennington 2013; Walter 2015; Walter et al. 2012; Van Den Hoven et al. 2008). However, other frameworks are also useful: Brubaker et al. suggested the DPM is a more 'clearly defined and operationalizable' theory when researching online memorials. Other researchers seem to agree: in their 2015 paper 'Death ends a life *not* a relationship: Timework and ritualization at Mindet.dk', Christensen and Sandvik provide an example of the DPM 'oscillation' being played out in the digital environment of the Internet (although they do not mention the

DPM by name). They suggest that Mindet.dk—a Danish website used by the parents of children who die very young or who are stillborn—creates a community of mourners. This space, they suggest, ritualises grief work through performances such as lighting a candle and writing messages to the child they have lost. They suggest this 'parallel existence' enables the parents to move on with their lives whilst maintaining a relationship with their dead child by 'continuously keeping hold' (Walter in Christensen and Sandvik 2015). There is an overlap of theories in the research which demonstrates the complexities around grief and bereavement, as although their research illustrates the DPM in practice, the continuity of the relationships still echoes the theory of continuing bonds.

## Off the Beaten Track: Growing Around Grief

The theory 'growing around grief 'is absent from much of the academic death studies literature. In a workshop run by grief counsellor and writer Lois Tonkin, a woman described a model which seemed to encapsulate how she experienced grief following the death of a child (Tonkin, Lois TTC 1996). The woman drew a circle and shaded it to represent how grief completely consumed her. She then explained that rather than grief shrinking over time, her life grew around the grief—this is sometimes described as the fried egg model. In practice, this growing around grief model seems to resonate with others who feel grief is not something to be worked through until a resolution is reached: instead, the grief becomes part of you, and you move on *with* grief, not *from* grief. Tonkin explained that for some, it 'explains the dark days' (1996, p. 10)—what I described earlier in this chapter as the unexpected 'wave' of grief. This valuable model of grief illustrates how grief and attachment to the deceased remain, but the world around them grows over time (Lloyd 2017). Most would agree that the death of a loved one fundamentally affects the lives of the bereaved as they attempt to adapt to their changed lives, and participant interviews reflected how grief is not something to 'get over' or something you can 'move on' from. This book is firmly contextualised within the theories of continuing bonds (Klass et al. 1996) and growing around grief (Tonkin 1996); however, I wanted to remain open to the potential that the proliferation of digital data—by and about the dead—may result in the need for a new framework, or may be an expansion of the existing frameworks.

## Digital Dislocation: The Disturbance to the Dual Process Model

Since Stroebe and Schut presented their DPM in 1999, it has been hugely influential in assisting bereavement counsellors, academics and professionals. As stated earlier, the model illustrates and explains what they suggest as healthy grieving, which involves an oscillation between 'loss-oriented grief' and 'restoration-oriented grief'—a movement between confronting and avoiding grief. However, the findings from the DI participants—presented in chap. 5—shed light on how posthumous digital inheritance can disrupt this oscillation between these two states by disturbing what Stroebe and Shut describe as the 'waxing and waning'. This disturbance, I suggest, creates a 'dislocation' for the bereaved not accommodated within the original model. Writing in 1999, Stroebe and Schut suggest that 'At times the bereaved will be confronted by their loss, at other times they will avoid memories, be distracted'; however, digital afterlives can stop this 'distraction' as the dead can confront us in new ways, as they reside in our laptops and in our pockets—they exist within our everyday devices.

The original psychological model of bereavement has been widely adopted across many different disciplines, including sociology. Nevertheless, I wanted to incorporate a sociological element within the model, which could be helpful when examining digital afterlives—particularly those created on social networks. Rather than adopt the psychological term 'stressor' for the additional element to the model, I adopted the more sociological term 'process' to illustrate that the digital-oriented sphere involves social processes and symbolic interactions reflecting the epistemological assumptions of this qualitative research.

Figure 6.2 shows the expanded DPM, which introduces the third sociological dimension—a digital-oriented process. This expanded model of coping with bereavement also suggests a process of 'flux', which not only encapsulates the complexity, uncertainty and uniqueness of how people experience grief, but it also illustrates the notion that—within and around the original two dimensions—the bereaved move messily rather than in straight lines of oscillation. Furthermore, adding this digitally oriented process to the model acknowledges the possibility that rather than being discrete, both original stressors can overlap and therefore can be experienced simultaneously. For instance, digital grievers can experience grief work *and* new roles/identities/relationships at the same time, which can then be accommodated within the new third, digital dimension. Obviously, digital grieving was not accounted for in the original model as it was developed in 1999 when the World Wide Web was still in its infancy, and posthumous digital inheritance was not possible. Therefore, an expansion to the model is necessary to highlight and acknowledge digital grieving and its effects on the process of oscillation—which is a fundamental element of the original DPM. Stroebe and Schut's model also focuses on the importance of distraction and the need to take respite from the original two stressors. However, findings of this research suggested that for some, this 'respite gap' is being partly occupied by the new digital-oriented process when the bereaved spend time with the socially active dead on their digital devices. I suggest that for some, this digital-oriented process not only *occupies* part of the 'respite gap' but—due to the disturbance to the oscillation—could cause a ***digital dislocation*** where the bereaved could become 'stuck' in what for some could be a complex labyrinth—a digital maze—of digital inheritance. The original model also implies that the bereaved have an element of control over the timing and speed—the waxing and waning—of oscillation between the loss-oriented and the restoration-oriented activities; however, as Kasket (2019) points out, these oscillations can be 'utterly unpredictable and uneven'. During data collection, I explored the frequency and triggers of visiting digital memories and messages, and the importance of control was implicit in responses from many of the DI participants (see Chap. 5). However, the delivery of timed posthumous messages would disrupt the ability to control the frequency and timings and being able to choose when to visit the dead is essential here; the dead choosing to visit *you* could interrupt the necessary oscillation; moreover, the anticipation of further future messages could result in feelings of 'stuckness' within the digital maze.

The new expanded model also includes three new 'digital tasks'. Firstly, the fear of second loss, which, as discussed earlier in this chapter, is the anticipatory fear of losing the data created by—or commemorating—the dead. Secondly, spending time with the socially active dead, because even though the dead no longer evoke notions of the uncanny valley (see Chap. 3), the fact that the dead are accessible on our everyday devices contributes to a disturbance to the necessary respite and oscillation which is crucial to the original DPM. Thirdly, the creation of digital bonds (discussed below) fits into both existing dimensions as it involves 'doing new things' and 'continuing' or 'relocating' bonds. However, the research findings demonstrated how digital memories and messages are experienced in unique ways by the bereaved. Furthermore, these three new digital tasks

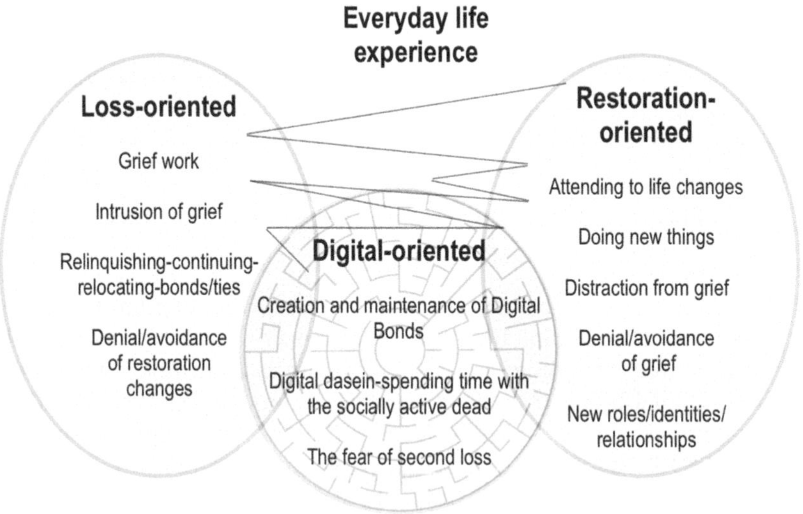


**Fig. 6.2** Expanded model of the Dual Process Model of bereavement

reflect how, for some, this type of data contains the essence of the dead, and for these reasons, I argue digital grieving should be treated as a unique type of grieving that necessitates an expansion of the DPM.

In 2016, Stroebe and Schut addressed what they identified as a 'major shortcoming' in their original model. They acknowledged that it did not address the issue of stressor 'overload', which they described as 'the bereaved person's perception of having more than s/he feels able to deal with – too much or too many activities, events, experiences and other stimuli' (Stroebe and Schut 2010). By suggesting an expansion of the DPM, I am not implying that digital grieving creates an overload that is *more* than people can deal with. Nor am I suggesting that digital-oriented grief is negative or unhealthy—I am suggesting it is a new and different digital 'griefscape' not accommodated within the original model. Furthermore, when the bereaved do not control the digital memories and messages, a warp-speed oscillation between the two original stressors could occur. As the research showed, those who possess control of digital memories and messages find them great comfort. Therefore, this suggested expansion should not be seen as an illustration of negative or harmful grieving; instead, incorporating this sociological dimension into the taxonomy of the DPM will make visible the impact of digital inheritance, and add to our current understanding of grief and bereavement. Echoing the sentiments of Stroebe and Schut (2010) this section of the chapter is an invitation for academics, professionals and counsellors, to theoretically and empirically test this suggested extension to assess its validity fully.

## Digital Bonds: An Expansion of Continuing Bonds Theory

Thanatechnology has changed the way people nurture continuing bonds with the dead. Continuing bonds provided a major theoretical framework for this research, and using bereavement theory as an anchor, I developed the term digital bonds to highlight and accommodate the unique nature of posthumous digital afterlives. The use of the term 'linking objects' (Volkan 1972) is helpful in discussions about continuing bonds as it describes how the bereaved use physical objects as tools to 'evoke the presence of the dead' thus enabling a continuing relationship. However, 'linking objects' does not adequately describe the universal nature of posthumous digital afterlives, which are often experienced differently from physical objects. We now carry precious digital memories and messages of the dead on our everyday devices, making it easier than ever to feel close to the dead and to continue a dynamic and ongoing relationship with them.

Continuing bonds theory describes the reconstruction of a person's social identity once someone in their lives has died (Klass 2006). Ongoing digital relationships—enabled by thanatechnology—differ from the one-way relationships accommodated within the original theory of continuing bonds. Using the term digital bonds highlights the public nature of posthumous digital memories and messages as it acknowledges that the dead are not passive in the creation of such bonds. By expanding continuing bonds and accepting digital bonds as a distinct subset, we recognise the creative input of accidental and intentional DCs who—along with DIs—operate as co-producers of these digitally facilitated bonds. I suggest these co-created bonds are then used and maintained in a digital *Dasein* allowing DIs to continue to have a relationship with the deceased. Notions of 'persons' and 'selves' then become less important because of the specific way this digital *Dasein* is experienced by the bereaved—as a unique place to spend time with the dead. Returning to the research findings, I argue that the emotional bonds accommodated within the existing framework cannot be lost by the DIs—notwithstanding the effects of memory loss—as they are the inner representations of the deceased, created by the bereaved. Digital bonds are *also* an inner representation of the deceased; however, DIs need thanatechnology to gain access which makes them different. In addition, digital bonds differ from traditional continuing bonds, as many of these technology-enabled bonds are created and maintained in the same place and space that the deceased used to create their original digital memories and messages (Kasket 2012). The theory of the fear of second loss developed from this research demonstrates how—following biological death—the anticipated social death of a person through technical obsolescence, accidental deletion or lack of access to these precious memories and messages could create anxiety for the bereaved who fear a severing of these precious digital bonds. Crucially here, this loss would not affect other ways of continuing a relationship with the deceased but again highlights the unique nature of digital afterlives and the need for digital bonds to be recognised as a distinct type of posthumous bond.

During the development phase of digital bonds theory, another new theory emerged from Oliver Clabburn at Edge Hill University, which I considered a useful and valid addition to continuing bonds. Clabburn developed 'The Model of Reciprocal Bonds Theory' (Clabburn 2018). Clabburn's research aimed to explore the experiences of how people affected by motor neurone disease (MND) experienced digital legacies, and although the research focused on one distinct group of DCs—those with MND—the idea of the *reciprocal creation* of bonds resonated with the idea of digital bonds theory. Moreover, Clabburn's theme of 'Conveyance of identity and past experiences' echoes the focused codes of image control that I coded under the overarching concept of 'quest for immortality': legacy creation; advise/inform future generation and story sharing. Equally, Clabburn's finding also echoed the findings from the DCs and the theme of 'therapy' as a motivation for the creation of digital memories and messages. Finally, it seemed that this research identified the same limitation to existing bereavement models: they do not take into account the creative input from the DCs (Clabburn 2018). This digital input from the DCs could influence the durable biography (Walter 1996)—created by the DIs—by introducing what Clabburn describes as an 'autobiographical chapter'. Clabburn's term 'autobiographical chapter' is a useful addition to the theory of continuing bonds, as it acknowledges the input from DCs who had created video digital legacies, in an effort to 'continue parenting' after their death (Clabburn 2018). The concept of digital bonds demonstrates how these durable biographies and autobiographical chapters are brought to life, or crucially brought *back* to life, by resurrecting and creating a more immersive way to remain connected to the dead.

Reciprocal bonds theory certainly allows for the reciprocal nature of digital memories and messages, and it also acknowledges how DCs—who have created them intentionally—play a part in the creation of continuing bonds after they die. However, the theory still does not accommodate non-reciprocal, accidental everyday posthumous digital memories and messages, which this study found to be experienced as precious treasures by many of the bereaved. As discussed in Chap. 5, it is the everydayness of accidental digital memories and messages that the bereaved take comfort in and repurpose for their needs. Blustein's (2008) notion of 'rescue from insignificance' is helpful to discuss the 'accidental' digital memories not purposefully created to exist in perpetuity. Blustein's approach resonates with the findings of this research: it is the everyday, mundane messages that participants wanted to 'rescue from insignificance' and which renders them treasured possessions.

Consequently, these messages take on a new and different role and become a specific type of continuing bond. Wallace et al.'s notion of ongoingness (2018) is applicable here as it acknowledges a continuing relationship with the deceased but crucially it allows for the relationship with the deceased to 'grow'. This growth and the ability to form new bonds using thanatechnology is a fundamental element of the concept of digital bonds theory. As with the suggested expansion of the DPM, the introduction of digital bonds theory is an attempt to 'keep the conversation moving' (Klass 2006) by highlighting the distinctive way the bereaved are using technology and digital afterlives to continue and nurture their relationships with the dead.

In this chapter we have looked at why digital afterlives matter, and the possible ramifications of not considering the notion of second loss or indeed the 'fear' of second loss. In the next chapter we will see how the DAI—both intentional and accidental—could mitigate potential problems for those who inherit, adopt or co-create digital memories and messages by adopting a variety of thanatosensitive design features (Massimi 2012) and a voluntary code of conduct.

## References

Becker, E. 2014. *The Denial of Death*. 3rd ed. London: Souvenir Press Ltd.

Blustein, Jeffrey. 2008. *The Moral Demands of Memory*. Cambridge University Press.

Bouc, Amanda, S. Hye Han, and Natalie Pennington. 2016. 'Why Are They Commenting on His Page?': Using Facebook Profile Pages to Continue Connections with the Deceased. *Computers in Human Behavior* 62: 635–643. https://doi.org/10.1016/j.chb.2016.04.027.

Bowlby, John. 1969. *Attachment and Loss*. Vol. 1. Basic Books.

Christensen, Dorthe Refslund, and Kjetil Sandvik. 2015. Death Ends a Life Not a Relationship: Timework and Ritualizations at Mindet. Dk. *New Review of Hypermedia and Multimedia* 21 (1–2): 57–71.

Clabburn, Oliver. 2018. Investigating the Use of Digital Legacies with People Affected by Motor Neurone Disease (MND): An Interpretative Phenomenological Analysis.

Degroot, Jocelyn M. 2012. Maintaining Relational Continuity with the Deceased on Facebook. *OMEGA-Journal of Death and Dying* 65 (3): 195–212.

Durkheim, E. 2001. *Suicide: A Study in Sociology*. London: Routledge.

Exley, Catherine. 2004. Review Article: The Sociology of Dying, Death and Bereavement. *Sociology of Health & Illness* 26 (1): 110–122. https://doi.org/10.1111/j.1467-9566.2004.00382.x.

Getty, Emily, Jessica Cobb, Meryl Gabeler, Christine Nelson, Ellis Weng, and J.T. Hancock. 2011. I Said Your Name in an Empty Room: Grieving and Continuing Bonds on Facebook. In *Proceedings of the 2011 Annual Conference on Human Factors in Computing Systems (CHI '11)*, 997–1000. https://doi.org/10.1145/1978942.1979091.

Gibbs, Martin, James Meese, Michael Arnold, and Bjorn Nansen. 2015. # Funeral and Instagram: Death, Social Media, and Platform Vernacular. *Information, Communication & Society* 18 (3): 255–268. https://doi.org/10.1080/1369118X.2014.987152.

Glaser, B.G., and A.L. Strauss. 1966. *Awareness of Dying*. Godin: In. Aldine Transaction.

Gray, Selina Ellis, and Paul Coulton. 2013. Living with the Dead: Emergent Post-Mortem Digital Curation and Creation Practices. In *Digital Legacy and Interaction*, 31–47. Springer.

Hjorth, Larissa, and Jung Moon. 2017. Visual Afterlife: Posthumous Camera Phone Practices. In *Selfie Citizenship*, 119–126. Springer.

Kasket, Elaine. 2012. Continuing Bonds in the Age of Social Networking: Facebook as a Modern-Day Medium. *Bereavement Care* 31 (2): 62–69. https://doi.org/10.1080/02682621.2012.710493.

———. 2019. *All the Ghosts in the Machine*. Robinson.

Kastenbaum, R.J. 2015. *Death, Society, and Human Experience*. 11th ed. Abingdon: Routledge.
Klass, Dennis. 2006. Continuing Conversation about Continuing Bonds. *Death Studies* 30 (9): 843–858.
Klass, Dennis, Phyllis R. Silverman, and Steven Nickman. 1996. *Continuing Bonds: New Understandings of Grief*. Philadelphia: Taylor & Francis.
Kubler-Ross, Elisabeth. 2009. *On Death and Dying: What the Dying Have to Teach Doctors, Nurses, Clergy and Their Own Families*. Abingdon: Taylor & Francis.
Lloyd, Caroline. 2017. *Grief Demystified: An Introduction*. Jessica Kingsley Publishers.
Lofland, J., and L. Lofland. 1995. *Analyzing Social Settings: A Guide to Qualitative Observation and Analysis*. 3rd ed. Belmont: Wadsworth.
Massimi, Michael. 2012. Thanatosensitively Designed Technologies for Bereavement Support.
Massimi, M., and R. Baecker. 2010. A Death in the Family: Opportunities for Designing Technologies for the Bereaved. In *Proceedings of the 28th International Conference on Human Factors in Computing Systems*, 1821–1830. https://doi.org/10.1145/1753326.1753600.
Moncur, Wendy, and Miriam Julius. 2014. *Story Shell: The Participatory Design of a Bespoke Digital Memorial*, 1–9.
Neimeyer, Robert A. 2001. *Meaning Reconstruction & the Experience of Loss*. American Psychological Association.
Odom, William, Richard Harper, Abigail Sellen, David Kirk, and Richard Banks. 2010. Passing on & Putting to Rest. In *Proceedings of the 28th International Conference on Human Factors in Computing Systems – CHI '10*, 1831. https://doi.org/10.1145/1753326.1753601.
Öhman, Carl, and Luciano Floridi. 2018. An Ethical Framework for the Digital Afterlife Industry. *Nature Human Behaviour* 2 (5): 318–320. https://doi.org/10.1038/s41562-018-0335-2.
Pennington, Natalie. 2013. You Don't De-Friend the Dead: An Analysis of Grief Communication by College Students Through Facebook Profiles. *Death Studies* 37 (July 2015): 617–635. https://doi.org/10.1080/07481187.2012.673536.
Schut, Margaret, and Henk Stroebe. 1999. The Dual Process Model of Coping with Bereavement: Rationale and Description. *Death Studies* 23 (3): 197–224. https://doi.org/10.1080/074811899201046.
Small, N. 2001. Theories of Grief: A Critical Review. In *Grief, Mourning and Death Rituals*, ed. J. Hockey, J. Katz, and N. Small. Buckingham: Open University Press.
Steinbeck, John. 2008. *The Winter of Our Discontent*. Penguin.
Stokes, Patrick. 2012. Ghosts in the Machine: Do the Dead Live on in Facebook? *Philosophy & Technology* 25 (3): 363–379.

———. 2015. Deletion as Second Death: The Moral Status of Digital Remains. *Ethics and Information Technology* 17 (4): 237–248. https://doi.org/10.1007/s10676-015-9379-4.

———. 2021. *Digital Souls: A Philosophy of Online Death*. Bloomsbury Academic.

Stroebe, Margaret, and Henk Schut. 2010. The Dual Process Model of Coping with Bereavement: A Decade On. *OMEGA-Journal of Death and Dying* 61 (4): 273–289.

Sudnow, D. 1967. *Passing On: The Social Organization of Dying*. Englewood Cliffs: Prentice-Hall.

Tonkin, Lois T.T.C., and Cert Counselling (NZ). 1996. Growing Around Grief—Another Way of Looking at Grief and Recovery. *Bereavement Care* 15 (1): 10. https://doi.org/10.1080/02682629608657376.

Van Den Hoven, Elise, Wina Smeenk, Hans Bilsen, Rob Simmermann, Simone de Wart, and Van Koen Turnhout. 2008. Communicating Commemoration. In *International Workshop on Social Interaction and Mundane Technologies (SIMTech '08)*.

Volkan, Vamik D. 1972. The Linking Objects of Pathological Mourners. *Archives of General Psychiatry* 27 (2): 215–221.

Wallace, Jayne, James Thomas, Derek Anderson, and Patrick Olivier. 2018. Mortality as Framed by Ongoingness in Digital Design. *Design Issue* 34 (1): 95–107. https://doi.org/10.1162/DESI.

Walter, Tony. 1994. *The Revival of Death*. Taylor & Francis.

———. 1996. A New Model of Grief: Bereavement and Biography.

———. 2015. The Dead Who Become Angels: Bereavement and Vernacular Religion. *OMEGA-Journal of Death and Dying*. https://doi.org/10.1177/0030222815575697.

Walter, Tony, Rachid Hourizi, Wendy Moncur, and Stacey Pitsillides. 2012. Does the Internet Change How We Die and Mourn? Overview and Analysis. *Omega: The Journal of Death and Dying* 64 (4): 275–302. https://doi.org/10.2190/OM.64.4.a.

CHAPTER 7

# The Future of Digital Death

*I mean, it's even batshit crazy I can even talk to you.* (Martha, Be Right Back, Brooker, 2012)

The Digital Afterlife Industry (DAI) is dynamic by its very nature: the ubiquitous use of smart technologies, the effect on interpersonal communication and the growing number of users ensure this industry will continue to grow at an ever-increasing rate. Research by Öhman and Watson (2019) suggests that by 2100 there will be 1.9bn deceased profiles on Facebook. With this in mind, it is unsurprising that the DAI is a fast-growing area of commerce. It is estimated that in 2018 the DAI in the United States was worth approximately US$16–20bn per annum, and an estimated £2bn per annum in the United Kingdom (Arnold et al. 2018). As we saw in Chap. 2, these immortality platforms are fragile: they continue to fail at a steady pace—a troubling thought when we consider the existential part they play in both the creation and the inheritance of digital afterlives. Issues around the fear of second loss discussed in the last chapter raise important philosophical questions. They are tangled up with issues surrounding the Right to Forget, the Right to Be Forgotten and the Right to Remember. Moreover, the notion of the fear of second loss draws attention to a crucial question in the study of digital afterlives: whose rights should be prioritised, the living or the dead? This penultimate Chapter is



D. J. Bassett, *The Creation and Inheritance of Digital Afterlives*, Palgrave Studies in the Future of Humanity and its Successors, https://doi.org/10.1007/978-3-030-91684-8_7

aimed at both accidental and intentional SPs and is a plea for them to consider the needs of the DIs (the living) alongside the technological and commercially driven needs of this dynamic and important Industry.

## Designing for Mourning in a Digital Society

Interestingly, the idea of designing for bereavement predates the prevalent use of SNS. In his 1995 paper, Edwin Bos predicted the emergence of platforms such as Eternime by discussing the possibility of extending life by using technology such as interactive talking photographs. He suggested that designing such technology needs to consider the bereaved, whom he suggested could be supported by such technology. This notion of designing for death was also discussed in a 1997 book by transhumanist Timothy Leary (see Gray 2015). Leary was terminally ill when he wrote his book, which suggested that systems could be designed to allow life events to be uploaded and recorded, which could then be used to immortalise your consciousness after death. This idea that material remains could play a part in the process of grieving, and how technology could be designed which would assist with this in a positive way, was taken up again at the SIMtech event at Microsoft Cambridge in 2008 (Foong and Kera 2008; van den Hoven et al. 2008; Kirk and Banks 2008).

The literature addressing distress at online expressions of grief plays an integral part in our ability to see the effect digital afterlives have on the bereaved. However, there is evidence that there has been a normalising of grief on the Internet (see Brubaker et al. 2013; Christensen and Sandvik 2015; Paulus and Varga 2015; Pitsillides, et al. 2013; Wagner 2018 Walter et al. 2012). However, this change in attitudes may also have something to do with the growing amount of research into what Massimi and Charise (2009) call thanatosensitive design (TSD), or 'death sensitivity'. In 2015, when I started writing my PhD thesis, many accidental digital memory platforms seemed to be playing catch-up with the idea that their users were using their platforms to discuss death, dying, and bereavement and retrieval and storage of digital memories of dead users. However, researchers and practitioners were beginning to discuss how platforms could be designed to be more sensitive to the bereaved (Brubaker and Callison-Burch 2016; Brubaker et al. 2013; Gray 2015; Massimi 2010; Massimi and Charise 2009). In 2009, Michael Massimi and fellow PhD student Andrea Charise argued for the use of death sensitivity—thanatosensitivity—when designing systems. They suggested that systems needed designing respectfully and compassionately, as they were used by people facing the end of their own life and also by those who were coping with the loss of a loved one. Digital objects or technological artefacts can help people 'overcome' loss (Massimi and Baecker 2010). As mentioned in the last chapter, words such as 'overcome' are unhelpful as they echo the idea of stage theory. However, Massimi and his colleagues' work is crucial to the sentiments of this book, as it is underpinned by the need to design systems with the bereaved in mind. Other researchers have also been looking at designing technology to help the bereaved remember the dead (Brubaker et al. 2013; Gray 2015; Getty et al. 2011; van den Hoven et al. 2012; Maciel and Pereira 2013; Mori et al. 2012; Odom et al. 2010).

## Designing *Without* Thanatosensitive Consideration

By 2012, research about the importance of appropriate TSD began to gain momentum, as stories began to emerge regarding the impact on the bereaved of designing *without* TSD. Facebook had faced much criticism for its lack of compassion when dealing with bereaved families. One instance, in particular, exemplifies this lack of compassion: in 2014, 20-year-old Hollie Gazzard was murdered by her ex-boyfriend, and at the time of her death, there were several photographs on Hollie's Facebook page of her posing with her murderer. Following her death, her family asked for the photographs to be removed. Facebook refused to take down the photographs, stating it was against their policy to delete photographs from memorialised profiles. For her 2019 book *All the Ghosts in the Machine*, Elaine Kasket interviewed Hollie's father, Nick, who explained how when he contacted Facebook to ask if specific photographs could be removed from the site, they suggested closing the profile completely. However, this was not what Nick or the rest of Hollie's family were trying to achieve—they wanted to be able to visit the site and post messages onto the page. Kasket described how it was only by using John Giacobbi 'The Web Sheriff' that the Gazzard family finally managed to get the photographs of Hollie's murderer removed from her Facebook page. As discussed earlier in this book, Facebook has made several changes to the way they deal with memorialised profiles. The ability to memorialise a Facebook account has been possible since 2007, and in 2015 Facebook introduced its Legacy Contact feature, which allows users to nominate a friend or family member to take 'stewardship' of their Facebook profiles after they die. In a 2016 paper, Brubaker and Callison-Burch set out the design choices that led to the development of the Legacy Contact feature.

> *Adopting stewardship as a framework enabled us to strike a balance between the needs of the account holder and the bereaved community.* (Brubaker and Callison-Burch 2016, p. 2)

Notably, the authors point out that 'design research' played a part in this new feature, demonstrating that platforms such as Facebook are using academic research from the various disciplines to inform how digital platforms can be designed with compassion and thanatosensitivity in mind, in an effort to better support the bereaved in a socio-technical society. The introduction of new Facebook features such as Tributes—discussed in Chap. 5—is yet another step towards TSD being incorporated into more SNS, and the forthcoming work led by Jed Brubaker at the University of Colorado could also contribute to this crucial area of thanatological research.

## Just Because You Can ...

Whilst I understand platforms exist that offer the delivery of timed posthumous messages and the digital reanimation of the dead, I am hopeful that this chapter will provide further insights and potential solutions to some of the issues around the creation and inheritance of digital afterlives. Few would argue that Moore's Law will result in the 'Black Mirror' technologies of the future being closer than we imagine, and this particular genie cannot be put back into the bottle. As we saw in Chap. 2, digital afterlife platforms are far from immortal; however, other types of platforms and tech companies fail all the time, so why should we care more about the DAI? Many of these platforms offer 'immortality'—not only to those doing the creation but those who inherit the precious data of the dead. Therefore, when these companies fail the ramifications for the bereaved are often profound. In Chap. 4, we saw how DCs who intentionally create posthumous digital messages to be delivered following their deaths do so for a number of reasons—none of which are intended to cause harm to DIs. However, how the platform is designed, and the information provided to DCs, is the responsibility of the SPs. Incorporating information for DCs highlighting potential problems for the bereaved caused by receiving uncontrollable posthumous messages could be a valuable and thanatosensitive addition to these platforms—which could, in return, become a sound commercial decision.

Messages recorded and stored on intentional posthumous sites to be delivered following the creator's death contain audio and are uncontrollable by the bereaved, which poses potential problems, as we saw in Chap. 5. Other recent research supports the notion that audio triggers different emotions to those triggered by other digital memories. Clabburn's 2018 study into the creation of digital legacies by those who suffer of MND and their subsequent inheritance following the death of the DC found that audio was 'the most emotionally evocative aspect of the recordings'. However, crucially Clabburn's research also found that these recordings provided comfort to some participants, even though they reminded them of their grief. Similarly, Brewer and Sparkes (2011) suggest that listening to the voice of their dead parent could help some children establish a continuing bond. Having the ability to *choose* when to visit this type of message is vital here; therefore, careful consideration should be given to those who inherit the messages *without* possessing any ability to control the timings. The knowledge that for the bereaved the choice of whether or when to listen to the voices of the dead is an important issue is even critical when—as this research shows—many SPs believe that the inheritor possesses complete control over whether to listen to their inherited messages once they are delivered: Ron (who owns a platform offering timed messages) told me how 'you can choose whether you want to view it or not, you will have to decide whether you want to go there and watch that movie', but as we saw in Chap. 5, even people who would not want to receive posthumous digital messages would feel compelled to listen (or watch) them if they knew of their existence.

Therefore, when a DI is informed that posthumous messages have been created for them, they should—if they so choose—have the ability to download immediately and hear (or watch) all messages that have been stored—even those that are timed for delivery on specific future dates. This would not only give control to DIs; it would also mitigate some of the potential data loss problems which would occur if the platform ceased to exist before the nominated delivery date. The possibility that messages which are to be delivered on future significant dates could be lost before DIs had the chance to listen to them—which the creator may not have anticipated—could cause problems for the bereaved by disrupting what is considered to be a healthy grieving process; furthermore, it could result in experiencing a second (digital) loss.

To address the issues around lack of control and thinking about possible solutions that the SPs could adopt, I began to hypothesise about the usefulness of a menu that would let the inheritor know the total number of messages created and the duration and suggested delivery dates. Numerous SPs had mentioned messages for 18th and 21st birthdays, wedding days, graduations and the birth of a child. Once I started contemplating how the menu could work, it highlighted one of the fundamental problems with timed messages: the assumption that these significant events are as profound to the DIs as they were to the DCs. Moreover, any menu would also need to consider social and cultural complexities that deal with family structures and relationships. With these potential problems in mind, I started to contemplate a menu that could offer more flexibility.

The idea presented in Table 7.1 is that the message creator could enter suggested dates of delivery that would be visible to the inheritor; however, the inheritor would have the ability to listen to all messages at a time of their choosing. Giving control to the recipient of the messages—rather than the creator—raises important ethical implications of consent. However, incorporating thanatosensitive design principles at the time of creation—as suggested by Massimi and Charise (2009)—and encouraging SPs to inform DCs about potential negative impacts for recipients, along with the introduction of control menus would be welcome additions to digital afterlife platforms. Although none of the participants in this research encountered digital messages which could be considered threatening or coercive, several people mentioned the possibility of people using posthumous messaging platforms to digitally haunt, stalk or troll the bereaved from beyond the grave by creating menacing messages to be delivered following the death of the creator. As we have seen in Chap. 5, DIs find it difficult to ignore posthumous messages. Moreover, although this is a hypothetical point not encountered in this research, I suggest that the inability—on the part of the inheritor—to disregard messages could be a potential problem for digital afterlife platforms in the future. Incorporating a control menu could mitigate the possibility of posthumous stalking and the unwanted delivery of posthumous digital messages. Some would argue that the dead always could haunt the living; however, it is essential to acknowledge that these digital memories and messages are not hidden away in dusty shoeboxes kept in the attic: they are on people's everyday devices—carried around on smartphones, iPads, and laptops—they seep into the everyday lives of the living.

**Table 7.1** Sample control menu

| *Content* | *Duration* | *Listen now* |
|---|---|---|
| Milestone birthday message 1 | 15 minutes | ▷ |
| Milestone birthday message 2 | 11 minutes | ▷ |
| Celebration message | 12 minutes | ▷ |
| Changing family message | 12 minutes | ▷ |
| Milestone message 1 | 15 minutes | ▷ |
| Milestone message 2 | 14 minutes | ▷ |

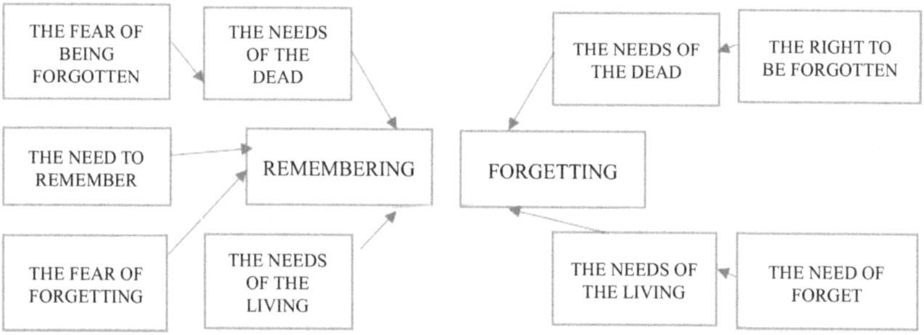


**Fig. 7.1** Remembering and forgetting the dead

## Remembering and Forgetting the Digital Dead

As we have seen, thanatechnology is changing *how* we remember the dead; however, *whether* and *when* we choose to remember the dead (which was once a private matter) has become entangled with the terms and conditions of some of the largest technology companies in the world. Human-computer interaction amplifies the fundamental tension between remembering and forgetting the dead in a digital society, where many of us unintentionally create digital afterlives through our everyday interactions on Apps and SNS. Figure 7.1 demonstrates the complex relationships between remembering and forgetting the dead.

Since the introduction of the General Data Protection Regulation (GDPR) in 2018, organisations are required to use 'clear and plain language' in their terms and conditions. However, currently, these terms and conditions do not legislate for the data of the dead. Legally, however, the dead do have certain rights: for example, they cannot be libelled but do have rights of reproductive liberty (Smolensky 2008). Whether there should be specific legislation for the data of the dead is a complex philosophical issue, because as we have seen, the data of the dead also include what I described in Chap. 1 as digital legacies and digital assets. Policymakers and lawmakers will be debating issues around privacy and rights on the Internet for many years to come, and it is outside the scope of this book to rigorously debate this issue. However, existing laws around privacy and rights do not fit comfortably in the digital realm when the data left behind are from a deceased user. Ongoing debates around the rights of the dead versus the rights of the living—such as whether the living have an obligation to remember the dead, along with ownership issues surrounding data of a deceased user—are complex and may never be fully addressed to everyone's satisfaction.

Earlier in Chap. 4, we saw how the 'quest for immortality' was the only theme coded from the posthumous message creator participants and how these messages could be seen as an attempt not to be forgotten following their deaths. Interestingly, none of these digital posthumous message creators coded under the theme of 'benevolence', which raises the question of whether these messages can be seen as a selfish act. One of the DI participants described how video messages recorded by her terminally ill husband brought her comfort because she could see how brave he was when facing imminent death—although she added that she still found it painful to watch the video and hear his voice. She explained how she talked to others about her husband every day and preferred this as a way to remember him rather than the video message he had created. This type of video message—created by someone who is terminally ill—seems fundamentally different to the type of timed messages being created by DC participants who are not dying: in that the need to be remembered seems less important than the need to say goodbye and bring comfort to those left behind.

Historically people have used the available technology to remember and memorialise the dead (see Wallace et al. 2018; Walter et al. 2012). The message creating DCs who took part in this research were looking for ways to be remembered after their biological deaths, but this begs the question of why they chose a digital format in this attempt at immortality. It could be due to the ease of the process; it could also be due to the accessibility of technology needed to create messages. However, I suggest that they may regard posthumous digital memories and messages as a permanent record. We have already seen that there are problems with this notion of the permanence of data stored on the Internet; however, many of the participants saw it as a way to be remembered in perpetuity.

Once again, listening to the voices of the dead seemed to be a dividing theme in the research findings. While some participants described the difficulty they found when listening to the audio, others explained they were frightened of forgetting the voices of their loved ones. For some, the desire to reconnect with their grief was an essential part of grieving, and audio seemed to be an effective way of facilitating this reconnection:

> *I like to watch them occasionally so I can be sure of remembering their voices. One of my fears when my parents died was not being able to remember their voices.* (Maggie)

For some participants, it was important to listen to the voices of the dead when they were having a good day, but for others, it was the opposite, once again illustrating the need to able to control when we chose to remember the dead. The ability to remember the dead is crucial and forms the very foundations of this book; however for some, the ability to forget the dead is also necessary. Over time photographs of deceased loved ones can be moved from room to room, house to house and sometimes into storage boxes in attics. However, with the digital dead, the need or desire for them to fade away is more complex. Do we 'delete' them or simply move them into a digital attic on a USB stick? Anyone who has made the difficult decision to delete the telephone number of a dead loved one from a device, only to have it re-appear from a storage cloud, knows how difficult this is.

Memories are fallible, and for those who suffer from dementia, digital technology can be a crucial aid for memory recall. While much of this book has focused on remembering and being remembered, the need to forget is also important for some. However, this is not a simple endeavour, as it is easier to collect and store digital data than attempt to delete it (Mayer-Schönberger 2011). Mayer-Schönberger argues there is a necessity to forget, but that 'technology facilitates the demise of forgetting'. This is an important point, and his notion that data files should have an expiry date and that users should understand and appreciate that information 'also has a lifespan' is an interesting one—once again, I am reminded of the idea of the 'forget' locket described by Wallace et al. (2018). Digital afterlives are making remembering and memorialising the dead easier than in the pre-Internet era, and in our digital society, there may be a need for a revival of forgetting (Mayer-Schönberger 2011). For the bereaved, choosing when and what they want to remember or forget—or just forget for the moment—comes back to the issue of control. It is improbable that the future dead intentionally want to harm the living by controlling how they grieve—even if they desire to be remembered in a certain way.

However, updating and manipulating the data of the dead—in a form of reanimation or digital resurrection—may not be what the creator intended. In the next section, we will consider some of the ramifications and complexities involved with 'tinkering' with the data of the dead.

## DDNR: Digital Do Not Reanimate

One of the aims of this book is to encourage SPs to inform users—who are intentionally creating digital afterlives in the form of avatars and thanabots—about the possibility that their posthumous data could be manipulated, which would result in them becoming digital zombies. It is essential that during the process of creating digital memories and messages, the creator should be encouraged to consider who will inherit their posthumous digital data and the effect this may have on them. In October 2020, Kanye West presented his then-wife Kim Kardashian with a hologram of her dead father for her 40th birthday. The digitally manipulated and resurrected Robert Kardashian wishes his daughter a happy birthday before reassuring her that she had met the 'most most most most most genius man in the whole world'. Kardashian described the hologram as a 'special surprise from heaven' and tweeted that she had been filled with emotion after watching the message. Robert Kardashian had become an accidental digital zombie, and there is no telling if he would have agreed to this type of digital resurrection. Both technology and the creator obviously mediated the hologram, and these types of digital inheritance and the emotions they evoke will be of ongoing interest to thanatologists—along with the motivation and potential manipulation behind such gifts when they laud praise on the bestower.

Early in the ethnographic research I began to understand how digital endurance seemed to blur the distinction between being dead and being alive, specifically when the deceased is biologically dead but still socially alive in a technologically mediated space. This blurring (once again, the idea of osmosis is helpful here) could, I suggest, eventually lead to a theoretical redefining of what it means to be dead. So far in this chapter we have considered the creation of both accidental and intentional digital afterlives and how the data of the dead could be manipulated to create digital zombies. However, is there something we could do to prevent our data being used in this way? In 2018, Öhman and Floridi recommended that SPs should guarantee 'users are not depicted radically differently from the bot that they originally signed up for'. Building on this idea, I suggest introducing a digital DNR. To elaborate, in end of life care a Do-Not-Resuscitate (DNR) order can be obtained from a doctor to instruct care providers not to administer CPR to a patient if their heart stops beating. However, from a digital afterlife perspective, the DNR initialism could be useful as an abbreviation for 'do not *reanimate*'. These digital DNR (DDNR) orders could mean that whilst someone does not mind their digital data becoming freely available to others following their death, they do not want it to be used to create any type of reanimated digital zombie (Bassett 2015); moreover, it would mean their posthumous avatar, thanabot or digital data could not be ideologically updated in any way. The suggestion here is that as part of avatar or thanabot creation, DAI platforms and Apps ask those doing the creation if they consent to future reanimation and, if the creator opted for a DDNR order, their creation would be unalterable following their death.

At the point of biological death, these DDNR orders can be considered the Schrödinger's cat moment of digital afterlives. When the creator dies, if a DDNR order is in place, there can be no reanimation of the data, and therefore no digital zombie is created. At this point, the digital version of the creator continues as the offline silicon version of the creator. This digital death would result in the data remaining static, unalterable, subject to decay, and finally, obsolescence: in other words, a *digital corpse*. However, if no DDNR is in place, the digital data *could* be reanimated, and the digital zombie is rendered rogue. This rogue digital zombie ceases to identify with its creator, ensuring the digital genes are free to evolve. Here it is worth clarifying that in suggesting the term 'digital corpse' I am referring to the *intentional* creation of thanabots and avatars rather than Stokes's 'digital remains', which I argue in Chap. 6 is not a useful term for accidental digital afterlives. In an attempt to ensure rogue digital zombies and thanabots do not outnumber the 'real' living, in the next section I suggest a code of conduct for the DAI, and these DDNR orders would form part of this voluntary code.

## Profit and Loss: A Voluntary Code of Conduct for the Digital Afterlife Industry

Whilst technology exists which facilitates the creation of digital afterlives, the policies and laws needed for this specific dimension of human-computer interaction will take time. At the outset of this thanatological research, the rights of the dead (DCs) versus the rights of the living (DIs) seemed a good place to start. However, as discussed earlier in this book, the term 'rights' places this discussion within a legal framework, and whilst the legal ramifications of digital afterlives are beyond the scope of this research, the moral ramifications form the cornerstone of this book. It is worth pointing out that many—if not most—users do not read the terms and conditions of their SNS accounts (Kasket 2019), and it can be assumed that most users are not thinking about their own mortality or digital inheritance when they sign up to these accounts. With this in mind, perhaps a voluntary code of conduct for the DAI would be a more suitable and practical tool, rather than waiting for complex legislation to be introduced.

Other researchers are calling for a similar framework: Carl Öhman and Luciano Floridi (2018) of the University of Oxford called for 'An ethical framework for the Digital Afterlife Industry' (DAI). They suggest that the DAI should be inspired by the ethical frameworks used by archaeological museums that deal with the monetising of organic human remains. A clarification of the relationship between DCs and SPs would, they claim, be helpful to ensure that the ownership of digital remains is decentralised (Öhman and Floridi 2018). They argue that digital remains hold an 'inherent value' and are not just data that can be monetised: this suggestion echoes the research findings, which uses the term 'essence' to describe the intrinsic value of digital remains. Although there are benefits of an ethical code or framework, as suggested by Öhman and Floridi, getting the DAI to adhere to it could be challenging, as it is likely to be difficult to enforce and would also take time to implement. There is also the problem of who would pay for its creation. With this in mind, I suggest that rather than trying to impose an ethical framework onto the DAI, a voluntary code of conduct may have a greater impact than trying to regulate the industry from the outside. As this research has demonstrated, intentional and accidental platforms provide useful tools for the bereaved; therefore, they have an essential role in developing any such framework. If a voluntary code was in place, then digital afterlife SPs may well see its adoption as a sound commercial decision that would not affect their market opportunities. Moreover, if it was co-created and also endorsed by grief councillors, thanatologists, bereavement professionals, and end-of-life caregivers, the SPs who adopt it could promote their voluntary membership on their platforms as a sign of their commitment to both the creators and the inheritors of digital afterlives.

The taxonomy introduced in Chap. 1 of this book—which draws attention to the intentionality of the DCs—could be helpful to this discussion around the implementation of a voluntary code of conduct, as it highlights the different forms of posthumous digital data and the different stakeholders involved.

Adopting the commercial term 'stakeholder' (Öhman and Floridi 2018), Fig. 7.2 illustrates that only one stakeholder exists for three of the four scenarios. It is only the intentionally created and appropriated digital memories and messages that have two stakeholders. This needs to be addressed by the intentional SPs, as currently, they give 100% of the control to the DCs without acknowledging the DIs as stakeholders, and as we have seen, ignoring them as stakeholders is both unhelpful and potentially harmful. In the diagram above, I indicate only one stakeholder for digital memories and messages accidentally created but not appropriated (SPs). An analogue scenario is helpful to demonstrate the reasoning behind this: while sorting through the physical belongings of a deceased friend or family member, you come across personal diaries or private letters. Do you decide to read, keep or destroy these items? Before the invention of the Internet, this scenario would involve personal or family decisions; however, many of these messages and memories are now stored on SNS or

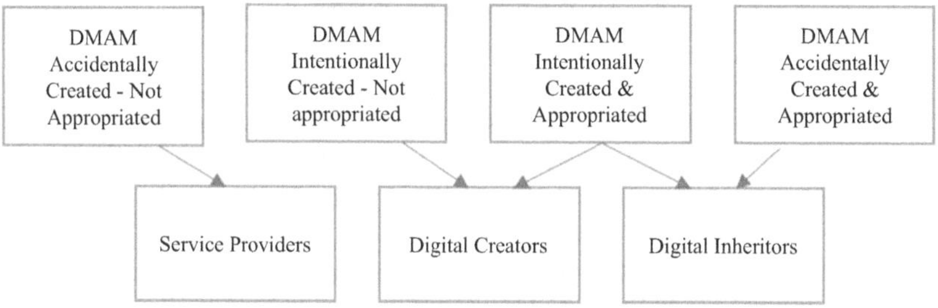


**Fig. 7.2** Stakeholders involved in the creation and inheritance of digital memories and messages

Apps in our digital society. These, once personal decisions are taken away from the bereaved and now regulated by global companies. When the memories and messages are ***accidental***, I argue any suggestions the online dead should be regarded as the only stakeholders (see Öhman and Floridi 2018; Stokes 2015) is flawed, as at the time of creation, the creator was unaware of—or had no interest in—future inheritance. Öhman and Floridi's (2018) suggestions focus on the dead and do not take into account the needs of the bereaved. For instance, they suggest that 'users only upload data that belongs to them personally, that is, not making bots out of a deceased relative or friend'; however, technology exists which enable people to make posthumous chatbots, so to envisage SPs putting this particular genie back in the bottle is naive.

Furthermore, as we have seen, although these posthumous chatbots can fall into the category of digital zombies (Bassett 2015), those creating and inheriting these forms of digital memories and messages can find them a comfort. Whether DIs should be denied this form of digital creation—as implied by Öhman and Floridi—once again comes down to the issue of control and who are considered to be stakeholders. I am not suggesting that the needs of DCs and SPs should be ignored; however, shifting the focus towards DIs could provide a more balanced approach. The suggestion is that storage and retrieval of posthumous messages would be at the forefront of the voluntary code, and all those adopting the voluntary code would need to prove the precious data are stored and are retrievable for a minimum amount of time. The needs of the commercially driven SPs are addressed within the company terms and conditions (which need to be agreed to by their users); and the needs of DCs are the concern of researchers such as Öhman, Floridi and Stokes. However, I hope this book shines a light on the needs of DIs, who I believe require further support and acknowledgement from the DAI, as the technologies behind the creation of digital afterlives march forward at an accelerating pace.

The sector of the DAI dealing with the creation of posthumous avatars are of particular concern to this research because, as we have seen, many people still view these avatars as eerie and uncanny. Although people are getting used to the socially active dead popping up on SNS, for many, socially active posthumous avatars still fall into the uncanny valley (Mori 2012). The exponential growth of emerging technologies needed to deliver and support the 'Black Mirror' imaginings of uploading our consciousness into our very own digital doppelgänger is a real possibility in the not-too-distant future (see Bell and Gray 2000; Meese et al. 2015;

Rothblatt 2014). As technology and the ability to create these avatars becomes more sophisticated, the need to ensure the DAI has a code of conduct in place will be crucial, not only to protect DIs but to inform those doing the creation of the potential harms caused by uncontrollable reanimated digital zombies (Bassett 2015). Thinking about thanabots, in 2018, Öhman and Floridi recommended that SPs should guarantee 'users are not depicted radically differently from the bot that they originally signed up for'. Here the implementation of a Digital Do Not Reanimate order (DDNR)—discussed earlier in this chapter—could mitigate the potential problems caused by reanimated digital zombies (Bassett 2015). My suggestion is that as part of avatar or thanabot creation, platforms and Apps ask those doing the creation if they consent to future reanimation—which would allow their avatar or bot to 'evolve'—or if they prefer their creation to be unalterable via a DDNR order—the Schrödinger's cat moment of digital afterlives.

This book calls for a collaborative approach by grief professionals, councillors, thanatologists and platform designers to produce a set of guidelines to set a 'gold standard' in thanatosensitive design for the DAI, specifically for the intentional digital afterlife SPs. These companies may benefit from the introduction of a code of conduct, as the DCs may be more inclined to use SPs who demonstrate they are committed to the needs of the inheritors by operating under such a code. The technological capabilities of these platforms and the services they offer should not take priority over the wellbeing of the creators *or* the bereaved—you may recall how ETER9 questioned whether avatar creation was a good thing. In the techno-social sphere of digital afterlives, not addressing the needs of the DIs could be a crucial oversight by these companies because if the inheritance of the reanimated dead is experienced in a negative way, this commercially driven industry cannot succeed. As we have seen, many of the digital afterlife SPs who took part in this research had extremely personal reasons for starting their companies; however, ignoring the commercial needs of these companies is not a realistic option as many of these companies receive finance from profit-driven investors. The next few years offer challenges and opportunities for the DAI, and the ideas posited here join other voices calling for the DAI to design their platforms with TSD in mind (Brubaker et al. 2013; Brubaker and Callison-Burch. 2016; Clabburn 2018; Gray 2015; Kasket 2019, Massimi 2012; Massimi and Charise 2009; Pitsillides 2017). Data preservation, storage and retrieval, can never be guaranteed; however, adopting a voluntary code of conduct could be a way of confronting the issues of potential data loss and technical obsolesces. Furthermore, together with the introduction of control menus for timed posthumous message creation and DDNR orders, this voluntary code of conduct could ensure the future dead who use these platforms take comfort in the knowledge that providers are aware of potential problems with digital inheritance and that they are not creating something that could cause harm to those who inherit their creations. While I understand the code of conduct I am suggesting in this book is voluntary, I hope that the DAI will see it as a welcome addition to their platforms and a benefit to all stakeholders.

In an effort to mitigate potential problems with the inheritance of digital afterlives this chapter has offered some practical suggestions for the DAI, which I hope will provide a starting point for further discussion and debate. In the next chapter of this book I look back at some of the ethical challenges faced during the research, the people I have met along the way and how my own life experiences have influenced this research and therefore the findings presented in this book.

## References


Arnold, Michael, Martin Gibbs, Tamara Kohn, James Meese, and Bjorn Nansen. 2018. *Death and Digital Media*. Routledge.

Bassett, Debra. 2015. Who Wants to Live Forever? Living, Dying and Grieving in Our Digital Society. *Social Sciences* 4 (4): 1127–1139. https://doi.org/10.3390/socsci4041127.

Bell, Gordon, and Jim Gray. 2000. Digital Immortality. *Communications of the ACM* 44 (3): 28–30. https://doi.org/10.1145/365181.365182.

Brewer, Joanne D., and Andrew C. Sparkes. 2011. Young People Living with Parental Bereavement: Insights from an Ethnographic Study of a UK Childhood Bereavement Service. *Social Science and Medicine* 72 (2): 283–290. https://doi.org/10.1016/j.socscimed.2010.10.032.

Brooker, Charlie. 2012. *Be Right Back. Black Mirror*. UK.

Brubaker, Jed R., and Vanessa Callison-Burch. 2016. Legacy Contact: Designing and Implementing Post-Mortem Stewardship at Facebook. In *Proceedings of the 2016 CHI Conference on Human Factors in Computing Systems, 2908–19*. ACM.

Brubaker, Jed R., Gillian R. Hayes, and Paul Dourish. 2013. Beyond the Grave: Facebook as a Site for the Expansion of Death and Mourning. *The Information Society* 29 (3): 152–163. https://doi.org/10.1080/01972243.2013.777300.

Christensen, Dorthe Refslund, and Kjetil Sandvik. 2015. Death Ends a Life Not a Relationship: Timework and Ritualizations at Mindet. Dk. *New Review of Hypermedia and Multimedia* 21 (1–2): 57–71.

Clabburn, Oliver. 2018. Investigating the Use of Digital Legacies with People Affected by Motor Neurone Disease (MND): An Interpretative Phenomenological Analysis.

Foong, P.S., and D. Kera. 2008. Applying Reflective Design to Digital Memorials. In *SIMTech '08 Workshop*. Cambridge.

Getty, Emily, Jessica Cobb, Meryl Gabeler, Christine Nelson, Ellis Weng, and J.T. Hancock. 2011. I Said Your Name in an Empty Room: Grieving and Continuing Bonds on Facebook. In *Proceedings of the 2011 Annual Conference on Human Factors in Computing Systems (CHI '11)*, 997–1000. https://doi.org/10.1145/1978942.1979091.

Gray, Selina Ellis. 2015. *Remains in the Network : Reconsidering Thanatosensitive Design in Loss*. Lancaster University.

Kasket, Elaine. 2019. *All the Ghosts in the Machine*. Robinson.

Kirk, D., and R. Banks. 2008. On the Design of Technology Heirlooms. In *International Workshop on Social Interaction and Mundane Technologies (SIMTech '08)*. Cambridge.

Maciel, Cristiano, and Vinicius Carvalho Pereira. 2013. Social Network Users' Religiosity and the Design of Post Mortem Aspects. In *Lecture Notes in Computer Science (Including Subseries Lecture Notes in Artificial Intelligence and Lecture Notes in Bioinformatics)*, 8119 LNCS, 640–657. https://doi.org/10.1007/978-3-642-40477-1_43.

Massimi, M. 2010. Technologies for Supporting Bereaved Families. In *Proc. CHI EA '10*, 2951–2954.

Massimi, Michael. 2012. "Thanatosensitively Designed Technologies for Bereavement Support."

Massimi, M., and R. Baecker. 2010. A Death in the Family: Opportunities for Designing Technologies for the Bereaved. In *Proceedings of the 28th International Conference on Human Factors in Computing Systems*, 1821–1830. https://doi.org/10.1145/1753326.1753600.

Massimi, Michael, and Andrea Charise. 2009. Dying, Death, and Mortality: Towards Thanatosensitivity in HCI. In *CHI '09 Extended Abstracts on Human Factors in Computing Systems*, 2459–2468. ACM.

Mayer-Schönberger, Viktor. 2011. *Delete: The Virtue of Forgetting in the Digital Age*. Princeton University Press.

Meese, James, Bjorn Nansen, Tamara Kohn, Michael Arnold, and Martin Gibbs. 2015. Posthumous Personhood and the Affordances of Digital Media. *Mortality* 6275 (February): 1–13. https://doi.org/10.1080/13576275.2015.1083724.

Mori, M. 2012. The Uncanny Valley. Trans K. F. MacDoram and N. Kageki. *Robotics and Automation Magazine* 19 (2): 98–100.

Mori, Joji, Martin Gibbs, Michael Arnold, Bjorn Nansen, and Tamara Kohn. 2012. Design Considerations for after Death. In *Proceedings of the 24th Australian Computer-Human Interaction Conference on – OzCHI '12*, 395–404. https://doi.org/10.1145/2414536.2414599.

Odom, William, Richard Harper, Abigail Sellen, David Kirk, and Richard Banks. 2010. Passing on & Putting to Rest. In *Proceedings of the 28th International Conference on Human Factors in Computing Systems – CHI '10*, 1831. https://doi.org/10.1145/1753326.1753601.

Öhman, Carl, and Luciano Floridi. 2018. An Ethical Framework for the Digital Afterlife Industry. *Nature Human Behaviour* 2 (5): 318–320. https://doi.org/10.1038/s41562-018-0335-2.

Öhman, Carl, and David Watson. 2019. Are the Dead Taking Over Facebook? A Big Data Approach to the Future of Death Online. June: 1–13. https://doi.org/10.1177/2053951719842540.

Paulus, Trena M., and Mary Alice Varga. 2015. 'Please Know That You Are Not Alone with Your Pain': Responses to Newcomer Posts in an Online Grief Support Forum. *Death Studies* 39 (10): 633–640. https://doi.org/10.1080/07481187.2015.1047060.

Pitsillides, Stacey. 2017. "Digital Death: The Materiality of Co-Crafted Legacies."

Pitsillides, S., M. Waller, and D. Fairfax. 2013. What Role Does Digital Information Play in the Way We Are (Re)Membered. In *Digital Death*, 75–90.

Rothblatt, M. 2014. *Virtually Human The Promise and the Peril of Digital Immortality*. 1st ed. New York: St. Martin's Press.

Smolensky, Kirsten Rabe. 2008. Rights of the Dead. *Hofstra Law Review* 37: 763.

Stokes, Patrick. 2015. Deletion as Second Death : The Moral Status of Digital Remains. *Ethics and Information Technology* 17 (4): 237–248. https://doi.org/10.1007/s10676-015-9379-4.

van den Hoven, Elise, Wina Smeenk, Hans Bilsen, Rob Simmermann, Simone de Wart, and Van Koen Turnhout. 2008. Communicating Commemoration. In *International Workshop on Social Interaction and Mundane Technologies (SIMTech '08)*.

van den Hoven, Elise, C. Sas, and Steve Whittaker. 2012. Introduction to This Special Issue on Designing for Personal Memories: Past, Present, and Future. *Human–Computer Interaction* 27 (May 2014): 1–12. https://doi.org/10.1080/07370024.2012.673451.

Wagner, Anna J.M. 2018. Do Not Click 'Like' When Somebody Has Died: The Role of Norms for Mourning Practices in Social Media. *Social Media + Society* 4 (1): 205630511774439. https://doi.org/10.1177/2056305117744392.

Wallace, Jayne, James Thomas, Derek Anderson, and Patrick Olivier. 2018. Mortality as Framed by Ongoingness in Digital Design. *Design Issue* 34 (1): 95–107. https://doi.org/10.1162/DESI.

Walter, Tony, Rachid Hourizi, Wendy Moncur, and Stacey Pitsillides. 2012. Does the Internet Change How We Die and Mourn? Overview and Analysis. *Omega: The Journal of Death and Dying* 64 (4): 275–302. https://doi.org/10.2190/OM.64.4.a.

CHAPTER 8

# Final Thoughts and Reflection

*As I hope becomes clear elsewhere Dad, I love you very much* (Nancy Sinatra)

The topics covered in this book are profound as they shine a light on some of the darkest moments of people's lives. The words 'data collection' do not adequately describe the time and emotion invested by participants in research into death and dying. Research into digital afterlives may sound impressive as an elevator pitch, with cutting-edge technology, avatars and thanabots involved. However, in reality, this technology impacts how people and society grieve. While some see the ability to create and inherit digital afterlives as a positive step that facilitates our continuing bonds with the dead, others see it as an existential challenge. In this final chapter, I reflect on some of the ethical challenges I faced during the research, and also, to give further context to the research findings, I reflect on the influence my life experiences had—and continue to have—on the research findings and the writing of this book.

## Self-reflection: Honesty and Learning

Qualitative research is an interactive process, which involves the researcher trying to explore the experiences and imaginings of the research participants in an attempt to understand the significance and meaning of how people experience a specific dimension of the social world (Mason 2012).



D. J. Bassett, *The Creation and Inheritance of Digital Afterlives*, Palgrave Studies in the Future of Humanity and its Successors, https://doi.org/10.1007/978-3-030-91684-8_8

Personal biographies often play an important part in the research project we choose (Lofland and Lofland 1995). This book is an exemplar of this idea as my biography influenced almost every aspect of this research. The decision to use constructivist grounded theory for the study was crucial, as this methodology enabled me as the researcher to acknowledge that this project was heavily influenced and embedded in my own life experiences. Critical self-reflection on biases and theoretical predispositions is essential (Urquhart 2013) because, without them, any piece of research is meaningless; moreover, research is never neutral and rarely comes out of nowhere—rather, it comes from within the researcher. When I was 23, my wonderful Dad took his own life. I knew I would never recover from this loss; never 'move on' from the loss and importantly a man I would I never 'let go' of. Long before I began this study, I began to question theories of grief that endorsed the idea that grief had stages that could—and should—be worked through (see Kubler-Ross 2009). Following my Dad's death, my Mother destroyed almost all photographs, letters and diaries containing anything personal from him or about him—although as an act of psychological bullying, she gave me his suicide note and death certificate as 'keepsakes'. The lack of any meaningful mementoes left a painful gap in the ability to recall his face or his voice. Therefore, when I became aware of technologies that enabled the reanimation of the dead, and the delivery of timed posthumous messages, it obviously made me think about my Dad and the note he left. I wondered how receiving this type of message on important days such as my wedding day or the birth of my children would have affected my grief. In the early research stages I decided not to recruit participants bereaved following the suicide of a loved one. I had a duty to myself and potential participants to ensure the research did not adversely affect my mental wellbeing. The exploration of whether thanatechnology has had any effect on the type of notes left—and how digital suicide notes may differ from physical suicide notes—could be significant and important bereavement research. Suicide notes are a physical form of posthumous messages and, therefore, when coding the interviews of the DC participants, I was aware that the idea of leaving timed posthumous messages was a problem for me personally: I was influenced by thoughts of how I would personally feel, and how this could negatively skew the interpretation and coding of these interviews. However, being aware of one's influence during the coding process is fundamental in qualitative research. Many thanatologists focus on bereavement following a suicide and I urge them to fill the gaps in the research presented in this book.

PhD research is a long process, and the emotional labour involved in an ethnographic study into death, dying and bereavement must be acknowledged (Hochschild 2012). I should also acknowledge that I have lost others I have loved; my brother Russell died 11 years ago aged 46, so for me—as for many others—death and grief are life experiences that have influenced who I am and how I live my life. Whilst I acknowledge that most books could be described as 'labours of love', it is important to acknowledge the specific needs of researchers who explore the sensitive and complex subject of death, dying and bereavement. In Chap. 1, I listed the terror attacks that have occurred during the writing of this book. Having to update this list throughout the numerous iterations was an emotional task that I found difficult. The emotional labour (Hochschild 2012) invested in thanatological research has been recognised by other 'sensitive topic' researchers (Gray 2015; Lee 1993; Massimi 2010; Moncur 2013). During this study, I kept a research diary which I found extremely cathartic following participant interviews. I also blogged on my website as a form of self-reflection. I was affected by each and every interview with the bereaved participants because I can empathise with their feeling of loss. However, I have also laughed with the participants, and learnt something about grieving from each and every one of them.

I have found the 'ongoing implications' of sensitive research to be both positive and negative; keeping a research diary certainly provided a creative outlet for the emotional response to the death of one participant. Burles (2017) notes that this reflective journaling can, in return, produce improved understanding and assist with finding meaning in negative outcomes: I have indeed found this to be true in this research. Outside support during the 'emotional labour' (Hochschild 2012) of thanatological studies is also crucial. Although I expected to use Twitter specifically as a recruitment tool, in the early stages of the fieldwork, I found it was being used by other PhD students and academics as a space to discuss research problems and offer advice and support. Being part of an online PhD researcher community on Twitter has been invaluable, and many of those mentioned in the acknowledgement pages are fellow tweeters. Sometimes research produces feelings of isolation, but the network of fellow scholars brought together via Twitter—and specifically fellow thanatologists—undoubtedly helped with these feelings.

Self-reflection is also invaluable, and when I took part in the podcast 'Death Goes Digital'—a podcast dedicated to open discussions around digital death—I realised once again how my life experiences had influenced the research: as an icebreaker for the interview, the editor asked for my favourite film, my favourite book and a song I would like to be played at my funeral. Without much thought, I picked *UP*—the Disney Pixar film; the book *Tuesdays with Morrie* by Mitch Albom; and the song 'What a Wonderful World' by Louis Armstrong. On reflection, I realised that both the film and the book's main focus was on how the dead continue to shape and contribute to the lives of those left behind. I had never been aware these two choices were connected—they were just my favourite book and film.

## #SAYTHEIRNAME

*I won't start crying but ... that makes me ...Gloria will be in your papers ... Gloria will be in your data*

Participant anonymity is a cornerstone of social research and forms a significant part of the general maxim of do no harm. When conducting research that confronts sensitive issues, special consideration needs to be given to anonymity. In this book, the names of all participants have been changed along with other identifiable features. Data collected for analysis from Twitter, thanablogs and Skype comments have been redacted to reduce the identifiability of the author. However, the Internet is an archived and searchable network, and there is evidence that even when data are anonymised an individual may still be identified (Fossheim et al. 2017; Markham and Buchanan 2012; Shklovski and Vertesi 2013; Zimmer 2010). On Twitter, I noticed how some people were using the hashtag 'saytheirname' to draw attention to how people avoid using the names of the dead. Hearing and seeing the name of their loved one brings comfort to the bereaved by ensuring their loved ones are remembered and by helping them maintain continuing bonds. During the research, three participants asked me *not* to change the names of the deceased in any published works which used their data. Moreover, one made it a condition of their participation. For these participants, taking part in the research presented in this book was an important act of remembrance. Challenges such as these typify the ethical complexities involved in bereavement research, because using the deceased's real names—who were not my research participants—could result in the identification of those who were my participants. Post-mortem privacy is an issue debated by other scholars (see Edwards and Harbinja 2013; Prates et al. 2015; Smolensky 2008; Sofka et al. 2012). However, as I wanted to stay within the 'spirit' of the ethical guidelines, and within the guidelines themselves, I needed to consider the possible ramifications of using real names when requested to do so. Changing the names of the deceased seemed disrespectful and dishonourable to the participants; moreover, it felt like an abuse of power in this qualitative study. With Mattingly's (2005) warning echoing in my ear that 'anonymity can turn to disrespect', I discussed the dilemma with other Internet researchers at various conferences and my PhD supervisors before making the decision. This research topic is not unique, and other researchers have faced similar ethical dilemmas (see Pitsillides 2017; Scarth 2016). In 2015, Walter discussed the problem of changing the names of the deceased when the deceased had been memorialised on the Internet. He suggests that when the bereaved have taken steps to memorialise the names of their loved ones online, then it seems unethical to remove the names for research purposes (Walter 2015). In addition, in a 2014 study into designing a digital memorial Moncur and Julius gave their participants the *choice* whether to have their name or those of their deceased loved ones changed. Treating the bereaved who participate in research as co-producers of knowledge—as discussed in Chap. 5—rather than a vulnerable group seems to be a more ethical approach. Pitsillides (2017) addressed this preconception of seeing bereavement as a vulnerable condition, as this clinical view she suggested could create an imbalance of power between participants and researchers. Scarth (2016) goes further by warning against what she calls 'forced anonymization'. Rather than seeing the anonymity issue as a problem or limitation of this study, here I challenge the traditional notion that a one-size-fits-all approach works when considering the anonymity of the dead by arguing participants are not passive social actors in this research, rather, I argue they are co-producers of knowledge (Phillips et al. 2013). Carmack and DeGroot (2014) noted an important issue about whether the benefits outweigh the costs—here I am talking about the benefits and costs to the participants—and the importance of research into thanatechnology and its impacts on society to the wider field of death and bereavement. Whilst never ignoring my responsibility to do no harm, I argue the Internet creates a blurring of the traditional dichotomous researcher-researched relationship because of what Marwick and Boyd describe as context collapse (2011). As discussed in Chap. 1, they use the term to describe how SNS 'flatten out' traditional notions of hierarchy for the bereaved; however, I think the term is helpful to describe how when used for data collection, the SNS blur and flatten the traditional imbalance of power between the researcher and participants. A consequence of this context collapse is discussed in the following section.

## When Participants 'out' Themselves

Following the publication of an article in *Discover Society*, during early 2017, I received feedback from several participants via Twitter about the last paragraph of the article in which I wrote:

> *When those involved in my research give their time and emotional energy to discuss their darkest moments and deepest grief, the use of terms such as 'participants' and 'preliminary findings' seems disrespectful. In this article I use the terms with sincere respect and thanks to all those who have given time for this study*

When I tweeted the paragraph, I described it as the most important part of the paper, and one participant replied publicly to the tweet:

> *And there she is in your digital lines, Thank you…*

This tweet appeared on the participant's Twitter feed, which clearly identified her as a participant in my research. This reply illustrates how participants see their loved ones in researcher's writing whether they are named or not. There were other instances where participants 'outed' themselves explicitly by commenting on my Twitter feed or implicitly by retweeting and liking my tweets. It should not be assumed that participants are embarrassed about taking part in this type of research. For many, it is quite the opposite—some participants expressed how they enjoyed being part of the research as it gave them chance to talk about their grief:

> *I got so excited after we connected on Twitter and agreed the interview. It will always stay as a deeply personal, meaningful and memorable experience with me* (follow-up mail correspondence with Amy)

> *I hope this is of some use to you, it was an unexpected opportunity for me to write (for the first time) how I felt about this picture and my relationship with it. Thank You for giving me that opportunity* (follow-up email correspondence with Jeremy)

These quotations support the suggestion by Dickson-Swift et al. (2006) that some participants find it therapeutic to participate in research, as it gives them the chance to talk about their deceased loved ones. We meet and engage with the bereaved as we go about our daily lives (Pitsillides, 2017), and as we saw in the last section, treating participants as passive agents is to treat them as 'vulnerable' people who need protection from their own decisions.

Others have unpicked the problematic issues around confidentiality and anonymity in qualitative research (see Scarth 2016; Tilley and Woodthorpe 2011; Walford 2005). However, we need to consider whose anonymity is being protected when we use pseudonyms for the dead, as the consequence of changing the names of the deceased raises ethical questions about whether we are adequately protecting the anonymity of our participants. In a 2016 study, Scarth examined the reasons participants wanted the real names of the deceased loved ones to be used. Furthermore, some of Scarth's participants had requested *their* names were not changed in the write-up of her study. From her findings, Scarth (2016) presented 3 themes (1) Symbolic immortality and memorialisation, (2) Openness: Death and dying and (3) Identity, ownership and respect. Considering these findings and reflecting on the similar ethical challenges experienced during this research, I decided not to change the names of the deceased *except* when requested to do so by my participants. To clarify, I am suggesting that the anonymisation of all names should be the default position, however, obtaining informed consent—the ethical baseline for all research—should not be the sole consideration for thanatological studies: allowing participants to choose whether they want the names of the deceased to be anonymised may be nearer the central tenant of sociological research—which is to do no harm. By the time this book was published, almost four years had passed since I had interviewed the three participants who had explicitly requested that I did not change the names of their loved ones. Therefore, just before sending the final draft to the publishers, I emailed each of the three to check if this was still the case. All three participants still felt the same way; therefore, throughout this book, the real names of their loved ones have been used.

## The Limitations

To my knowledge, this is the first study that takes a holistic look at digital memories and messages by looking at all three stakeholders in the production and inheritance of digital afterlives: the SPs, the DCs and the DIs. However, as with all studies, there are limitations to its scope and purpose. Firstly, it is important to acknowledge the socio-geographical limitations of this study; although I have discussed 'Western' mourning rituals throughout the book, there is a huge difference between acceptance and denial of death *within* the Western world: the UK, Ireland, Canada and the USA are each distinct from each other in their traditions, acceptance and denial of death and dying, and this book has taken a macro view of these differing English-speaking regions without comparing responses across these geographical boundaries. Moreover, throughout this book I have used the term digital society without any meaningful consideration of the digital divide—specifically the digital 'have-nots'. The lack of access to the Internet and the smart digital devices necessary to mediate our relationships with the dead creates groups that are then by definition absent from these research findings.

This interpretative study did not set out to represent all bereaved people who have inherited digital memories and messages; however, it did aim to be generalisable. From the onset, I had a firm understanding that grieving and bereavement are experienced in multiple and complex ways. However, this book is the first step towards a focused perspective on the impact of the creation and inheritance of digital afterlives. In order to strictly adhere to ethical guidelines, the experiences of participants using thanatechnology with complicated grief are outside the scope of the research. This would be a relevant and vital area for future studies, and a valuable addition to the findings presented here in this book.

Throughout this research I have teetered on the periphery of transhumanism, cyborgs and inforgs (Floridi 2014), however, by trying to remain focused on death, dying and bereavement—rather than what it will mean to be human in the future—these relevant and interesting areas of study are largely absent from the book. One of the main challenges of books that focus on digital afterlives has been keeping up to date with the fast-paced growth of this area of bereavement research. The exponential growth of the technologies involved is almost matched by the amount of research being done in this field. It will be no surprise that the proliferation of papers being published across multiple disciplines has been difficult to keep pace with. This has presented challenges and opportunities; some published research seemed to 'steal my thunder' or 'pip me to the post' by beating me to publication, and—if I am honest—negatively impacting my confidence as a researcher. Öhman and Floridi's 2018 work on providing a new framework for the DAI, and Clabburn's 2018 thesis, which introduced reciprocal bonds theory are just two examples. However, in hindsight, these research findings motivated me to develop the theories further, resulting in the suggested changes to the existing models of bereavement presented in this book.

A further limitation to the research was that I did not possess any clinical experience of grief and bereavement; however, this ensured I did not start the data collection weighed down with academic knowledge about past or current theories and models of bereavement. This lack of clinical experience suited the ontological position of grounded theory—that theory should be grounded in the data rather than generated by empirically testing existing theories. Furthermore, as Charmaz argues, constructivist grounded theory acknowledges '*how the researcher's interest, positions, and assumptions influenced his or her inquiry*' (2014, p.344). As noted earlier, I have been reflexive throughout the research and the writing of this book. In some ways this book tells my story as a researcher alongside the stories of my participants. Many participants commented that they found participating in this research therapeutic, and I experienced this too. Writing this book has allowed me to share an extremely personal part of my life—invisible to most—yet a significant part of who I am. It was an honour to spend time with my participants, and their stories have shaped this research and will stay with me forever. One final reflexive note: I would like to acknowledge that although I understood qualitative research is designed to elicit rich data, I truly underestimated how thoroughly emotional yet enjoyable it would be.

## A Final Word

As noted in the introduction, writing a book to inform and appeal to academics alongside business owners and technology designers was never going to be easy. However, I have endeavoured to write the chapters in such a way that it is possible for the various stakeholders to dip into the relevant sections of interest. While some researchers focus on the rights of those creating digital afterlives or the rights of the dead, to give balance to the ongoing debates about the digital persistence of data, this book makes no excuses for prioritising the experiences of those who inherit digital afterlives: the bereaved. Although sociological in nature, this book has looked beyond the disciplinary boundaries of sociology in an effort to encourage a multidiscipline, collaborative approach to this nascent and important area of research. I am hopeful that by writing this book, I am sending a metaphorical canary into the mine, which will result in *all* stakeholders—SPs, DCs and DIs—benefitting from the experiences shared by the participants who took part in this research.

Through these chapters, I hope I have made a convincing argument about the importance of digital afterlives. However, I have saved a poignant story until the end, as I want to leave you with a real sense of the importance of digital memories and messages to the bereaved. The data of the dead may well be made up of ones and zeros, but as I have demonstrated, for some, they are precious possessions that contain the essence of the dead. In 2018, Sally told me how she treasures a digital recording of her deceased daughter's heartbeat, stored on her mobile telephone. When the nursing staff called Sally to her daughter's bedside following a tragic accident, they asked her if she wanted to record her daughter's heartbeat on her smartphone before the machines keeping her alive were switched off. This story stuck with me: learning that this mother carried her dead daughter's heartbeat in her pocket and listened to it every night, perfectly captured why the research presented in this book is important: a heartbeat represents life, and even when it is digitally recorded and technologically mediated, it is far more than the ones and zeros of digital code.

The dead have played an essential part in this book; however, we should be aware that *we* are the future dead. For most of us who live in the digital societies of the West, technology is ensuring we will all have a digital afterlife. As we become more comfortable with the socially active dead being online, we should also be aware that we can lose these precious digital memories and messages containing the essence of our dead loved ones. The book is a cautionary tale; it reminds us that we are far from immortal. The impact and significance of thanatechnology should not be underestimated, yet even in the digital age, we would be wise to realise, we only live twice.

Finally, this book ends as it began, with the words of Terry Pratchett:

## Further Resources

The netiquette guide below has been reproduced here with the kind permission of Carla Sofka and should not be reproduced elsewhere without explicit permission from Carla Sofka (csofka@siena.edu).

**Netiquette Regarding Digital Legacies and Dealing with Death, Tragedy and Grief.**

**Compiled by Carla Sofka (csofka@siena.edu)* (See source list)**

### *Digital Legacies: Planning Ahead and at Time of Death*

1. Don't assume that you know what someone would want to happen to their digital assets and social media. Be proactive and learn about deceased user policies on social media (e.g. memorialising a FB page/Tribute section vs. creating a private FB group in memory of the deceased). Discuss someone's wishes regarding their digital legacy (www.digitallegacyassociation.com) and how they prefer to receive 'bad news' before a crisis or loss occurs.
2. NEVER request changes to the deceased's social media platforms or online services UNLESS you received permission from the deceased to act on their behalf (e.g. you are the assigned Legacy Contact or Inactive Account Manager).

### *Sharing 'Bad News'/Death Notification/Informing About Events*

3. 'Soften the blow' (Ganger, 2019): regardless of how you are using technology to announce a death, provide a warning that the message/posting contains sad news so that someone can choose to postpone reading it if the timing/circumstances warrant a delay.

4. Digital obituaries may be provided through the services of a funeral home. Coordinate the timing of its posting online with the funeral director to allow time for more personalised notifications. Once an obituary is online, control over the sharing of the news is lost.
5. Think before you text, post or tweet. Families need time to notify everyone who deserves to learn of an illness, death or tragedy in a personal manner. Ask yourself the question: Is it your story to tell? Where do I fall in the 'mourning hierarchy' (Walter, 2015)? If you are not a primary mourner (immediate family member or close significant other) or have not received permission from someone in the deceased's inner circle to share news of a death, it's not your right or responsibility to make the news public. If you have not seen postings about the death or an online obituary that has been shared on the social media of the primary mourners, this most likely reflects a conscious choice not to do so or a decision to wait.
6. DON'T ask questions about the tragedy/death on social media or expect details about the death to be posted publicly. Privacy is more important than curiosity; information may be excluded for a very good reason.
7. Once personal notifications of the death have taken place, many people choose to post details about a wake/visitation, funeral, or celebration of life IF they are public events. Using social media to share practical information can save time and energy. (Avoid sending Evites since they can go to spam and these events are time sensitive.) Do NOT share information about services or memorial events on social media WITHOUT PERMISSION from the family or other primary mourners.

### *Grieving Online*

8. Avoid posting vague comments that might cause someone to be worried about your well-being or safety.
9. Avoid 'comparing' your grief to others. There's no need to 'one up' or disenfranchise/invalidate another person's reaction to a loss. DON'T turn the loss into a competition (As Woods notes, it's not the 'Grief Olympics—see Hiss, n.d.)
10. Carefully consider where you choose to post information about your grief and the impact that your posts will have on others. If you

post about your grief on your personal social media, it will be on display for everyone who has access to that page.

a. Consider the benefits of creating a page with restricted access for those who knew the deceased. This provides people with more control over when they choose to expose themselves to information that may be emotionally intense. It also creates a layer of protection against memorial trolls/RIP-trolling.

b. Is it more appropriate to share messages privately (via Messenger or another non-public aspect of social media)?

c. Remember that if you tag the deceased in photos that they might show up in other family members' and friends' news feeds, which can be potentially unwelcomed and overwhelming.

11. DON'T feel guilty for unfriending someone who has died. If it's too painful to keep seeing them tagged in posts, do what is right for you.
12. DON'T impose your religious/spiritual beliefs and/or coping strategies on anyone else.
13. Be mindful of how your social media settings influence how easily others can share the information that you post. Regardless of whether or not your postings can be shared with one click, you cannot control what others do with the information that you post online.

### *Providing Support to the Bereaved*

14. Confirm that death notifications have taken place (see #3) before using social media to post messages of condolence and support.
15. Be respectful of differences in how people deal with illness, death, tragedy and grief. Some people are very private and believe that the use of social media and social networking sites (SNS) is inappropriate, while some are comfortable talking about their experiences and sharing emotions publicly. If someone is NOT using social media/SNS to grieve, support them in 'traditional' ways (send a card; talk with them in person or call on the phone (texting may be preferred by those who have grown up communicating this way; hugs if appropriate. For guidance about supporting the bereaved offline ('in real life—IRL), see the resource list below 'Resource List: Supporting the Bereaved Offline'.) Even if someone uses

social media during times of grief, providing support in real life is always a good idea, especially if YOU are not comfortable providing support online.

16. If someone is posting about their grief, acknowledge that you've seen it by using an appropriate reaction button/emoji and match the 'tone' of their post if you include a comment.

### *Providing Support to the Bereaved (Continued)*

17. If you perceive a discrepancy between how you think someone 'should be reacting or feeling' and what you observe, avoid criticising or judging these behaviours/responses. Reactions to grief vary across people and change with time. For example, it is normal and healthy to experience moments of lightheartedness and joy (and perhaps to post about this), even during times of intense grief.
18. Avoid posting 'clichés' or 'platitudes' (see the resource list below 'Resource List: Clichés/What NOT to Say').
19. Sharing memories or stories is appropriate and is often appreciated. Use good judgement about what is socially appropriate to post publicly on social media versus what should be shared privately.
20. Acknowledging a loss on significant dates in the future can provide comfort to the bereaved

    If someone has NOT posted about their grief on social media, do this PRIVATELY. Consider making a phone call, sending a 'thinking of you as you remember him/her' message (a card in the mail or a private message) on the deceased's birthday or other dates of remembrance such as the anniversary of the death or a significant event at which the deceased's presence will be missed (a graduation, wedding, Father's/Mother's Day, holidays).
21. DON'T speak ill of the dead or criticise the bereaved. If you can't say something nice, don't say anything.
22. DON'T expect a response. Grief is exhausting, and someone may not have the time, energy, or desire to read and respond to posts on social media. Remember—it's not about you!
23. Accept strangers who post respectfully—they are part of the grieving community.

    1. Some people who do not personally know the deceased or their family may want to reach out due to 'experiential empathy'—

the ability to relate to one's grief due to having experienced a similar loss. Grieving the death of a stranger may provide them with a socially acceptable outlet for their grief as well as the benefit of helping another griever feel less alone.

2. People who have never met IRL can form significant relationships online (e.g. virtual gaming communities, FB interest pages). Cyberloss isn't any less genuine simply because the interactions took place in cyberspace.

### *Taking and Sharing Images/Photos*

24. It is generally not a good idea to photoshop an image of the deceased or turn their image into a meme.
25. Funeral selfies are controversial. It is generally considered inappropriate to post a selfie that includes the deceased in the photo. For younger generations, selfies are not considered to be irreverent; they may simply be a way to acknowledge a loss or inform others about one's grief. Talk with young adults about the etiquette of taking photographs at a visitation/wake, funeral or gravesite before arriving. It may be considered offensive to post a 'check-in' at a funeral home or cemetery.

Please do not reproduce without explicit permission from Carla Sofka (csofka@siena.edu).

### *Taking and Sharing Images/Photos (Continued)*

26. DON'T sensationalise another person's tragedy. Sharing grim details, posting gruesome images or recordings (e.g. the scene of the death or accident, fire or shooting), or even posting the photo of someone's gravesite can be disturbing and painful, particularly if they appear in a social media feed when someone is not expecting it.

### *Live-streaming/Webcasting a Funeral or Memorial Service*

27. NEVER live-stream or webcast a funeral or memorial service without explicit permission from the primary mourners as well as any-

one who has a primary role in the service. Carefully consider how access to the live stream is gained (privacy settings on social media or if viewing is by invitation).

28. If there is an open casket (as noted in #25), including an image of the deceased is most likely not appropriate. To include an image of the deceased, consider using a favourite photograph of the person taken during life.
29. Guests should be made aware that the service will be live-streamed. The camera should not be focused on the people in attendance. Respect the privacy of the mourners who may not want their reactions during a potentially emotional time to be observed by others or captured on a recording. Check with the primary mourners and ask if they want footage included from any processional or recessional that would include them entering or leaving the service and respect their wishes.
30. Be mindful that there may be copyright issues if recorded music is being used in the service; check with the funeral director about whether or not it should be included.
31. If a recording of the funeral or memorial service is made, do not share it through any mechanism (public or private) without explicit permission from the primary mourners. They have the right to control how and with whom it is shared.

### *Digital Survivor Advocacy: Using Digital and Social Media Following a Death or Tragedy to Advocate for a Cause or a Change in Awareness, Behaviour or Policy (Also Known as Hashtag Activism)*

#32. If a primary mourner has already publicly commented on an aspect of the individual's life and/or death regarding a particular cause (a specific illness, suicide, overdose, etc.), it may be appropriate to honour that person's memory by doing the same in relation to an advocacy effort IF the primary mourner provides permission. (See #5. If you are not a primary mourner, it is not appropriate for you to disclose potentially private information about someone's cause of death or personal details about their life using a hashtag.)

#33. Be aware that the media may reach out to primary mourners or others who are participating in advocacy campaigns. Not everyone is comfortable becoming 'newsworthy', so respect a primary mourner's decision if they do not wish to 'go public'. Not everyone is comfortable participating in advocacy efforts; for those who make the decision to do so, the time following a death that is taken to be 'ready' varies widely.

#34. Respect the way that people have chosen to frame the cause or issue that is the focus of their advocacy efforts. DON'T substitute or propose a different word or phrase in a hashtag since doing so discounts the importance and relevance of the cause or issue that they have chosen to address.

#35. DON'T argue with or criticise someone if you do not agree with their beliefs and/or their point of view about a controversial issue. It is disrespectful and inappropriate to incite controversy on a social media posting or memorial site that was created to honour someone's memory. Post your personal beliefs or opinions on your own social media or in a separate digital space rather than imposing your beliefs in their space. For example, a friend posts on Facebook that they just found out that their cousin who is a police officer died in the line of duty and uses the hashtag #BlueLivesMatter. This post about their loss is not the place to publicly comment on your opinion about the hashtag.

Netiquette compiled by Carla Sofka using information from personal experience, her own research on social media and research, and the following sources:

Bartz, A., & Ehrlich, B. (2011, October 12). The dos and don'ts of online condolences. https://www.cnn.com/2011/10/12/tech/web/online-condolences-netiquette/index.html.

Cann, C.K. (2014, March 5). Grief online: The dos and don'ts of internet etiquette. Retrieved from https://www.huffpost.com/entry/grief-online-the-dos-and_b_4906291.

Ganger, B. (2019, January 11). How to announce a death on social media. Retrieved from https://beyondthedash.com/blog/how-to-announce-a-death-on-social-media/4951.

Sources (continued)

Clark, E. (2018, November 3). 10 guidelines for using social media to express grief. https://blog.thewellnessuniverse.com/10-guidelines-social-media-express-grief/

Lamothe, C. (2018, October 10). How to grieve for online friends you had never met in person. Retrieved from https://www.nytimes.com/2018/10/10/well/family/how-to-grieve-online-friends-you-never-met-in-person.html.

***Resource List: Supporting the Bereaved Offline***

Directive Communication Systems, Sofka, C., & Poskanzer, L. (2019, March 7) *Netiquette.*

*Experts Agree: It's Time for Guidelines on Death in a Digital World* [Press Release] Retrieved from https://www.directivecommunications.com/netiquette-experts-agree-guidelines-are-needed-on-death-in-a-digital-world/

Sabra, J.B. (2017). "I hate when they do that!". Netiquette in mourning and memorialization among Danish Facebook users. Journal of Broadcasting and Electronic Media, 61(1), 24–40.

Walter, T. (2015) New mourners, old mourners: online memorial culture as a chapter in the history of mourning, New Review of Hypermedia and Multimedia, 21:1–2, 10–24.

Athen, L. (n.d.). What not to say to a griever. Retrieved from http://www.griefspeaks.com/id9.html.

Smith, M., Robinson, L., & Segal, J. (2019, June). Helping someone who's grieving: https://www.helpguide.org/articles/grief/helping-someone-who-is-grieving.htm.

Healthbeat (n.d.). Ways to support someone who is in grief. Retrieved from https://www.health.harvard.edu/mind-and-mood/ways-to-support-someone-who-is-grieving.

Healgrief.org (n.d.). How to help someone grieving. Retrieved from https://healgrief.org/how-to-help-someone-grieving/.

Hoggart, A. (2018, January 26). 21 ways to help someone you love through grief. Retrieved from https://time.com/5118994/advice-for-helping-grieving-friend/

Mudlaff, S.J. (n.d.). Avoiding the clichés of grief. Retrieved from https://www.mcgill.ca/counselling/files/counselling/supporting_the_grieving_-_avoiding_the_cliches_of_grief_0.pdf.

### *Resource List: Clichés/What NOT to Say*

Athen, L. (n.d.). What not to say to a griever. Retrieved from http://www.griefspeaks.com/id9.html.

Healgrief.org (n.d.). How to help someone grieving. Retrieved from https://healgrief.org/how-to-help-someone-grieving/.

Mudlaff, S.J. (n.d.). Avoiding the clichés of grief. Retrieved from https://www.mcgill.ca/counselling/files/counselling/supporting_the_grieving_-_avoiding_the_cliches_of_

## References

Burles, Meridith Clare. 2017. Negotiating Post-Research Encounters: Reflections on Learning of Participant Deaths Following a Qualitative Study. *Mortality* 22 (2): 170–180. https://doi.org/10.1080/13576275.2017.1291605.

Carmack, Heather, and Jocelyn M. Degroot. 2014. Exploiting Loss?: Ethical Considerations, Boundaries, and Opportunities for the Study of Death and Grief Online. *Omega: Journal of Death and Dying* 69 (4): 315–335. https://doi.org/10.2190/OM.68.4.b.

Charmaz, Kathy. 2014. *Constructing Grounded Theory*. Sage.

Sofka, C., I. Cupit, and K. Gilbert. 2012. In *Dying, Death, and Grief in an Online Universe*, ed. Carla J. Sofka, Illene N. Cupit, and Kathleen R. Gilbert. New York: Springer Publishing Company.

Dickson-Swift, Virginia, Erica L. James, Sandra Kippen, and Pranee Liamputtong. 2006. Blurring Boundaries in Qualitative Health Research on Sensitive Topics. *Qualitative Health Research* 16 (6): 853–871.

Edwards, Lilian, and Edina Harbinja. 2013. 'What Happens to My Facebook Profile When I Die?': Legal Issues Around Transmission of Digital Assets on Death. In *Digital Legacy and Interaction*, 115–144. Cham: Springer.

Floridi, Luciano. 2014. *The Fourth Revolution: How the Infosphere Is Reshaping Human Reality*. Oxford: OUP.

Fossheim, Hallvard, Helene Ingierd, Dag Elgesem, Charles Ess, Anders Olof Larsson, Marika Lüders, Robindra Prabhu, Katrine Utaker Segadal, Elisabeth Staksrud, and Kari Steen-Johnsen. 2017. Internet Research Ethics.

Gray, Selina Ellis. 2015. *Remains in the Network: Reconsidering Thanatosensitive Design in Loss*. UK: Lancaster University.

Hochschild, A.R. 2012. *The Managed Heart: Commercialization of Human Feelings*. California: University of California Press.

Kubler-Ross, Elisabeth. 2009. *On Death and Dying: What the Dying Have to Teach Doctors, Nurses, Clergy and Their Own Families*. Abingdon: Taylor & Francis.

Lee, R.M. 1993. *Doing Research on Sensitive Topics.* Sage.
Lofland, J., and L. Lofland. 1995. *Analyzing Social Settings: A Guide to Qualitative Observation and Analysis.* 3rd ed. Belmont: Wadsworth.
Markham, Annette, and Elizabeth Buchanan. 2012. Ethical Decision-Making and Internet Research Recommendations from the AoIR Ethics Working Committee. *Recommendations from the AoIR Ethics Working Committee (Version 2.0)*, 19. Retrieved from www.aoir.org.
Marwick, a. E., and D. Boyd. 2011. I Tweet Honestly, I Tweet Passionately: Twitter Users, Context Collapse, and the Imagined Audience. *New Media & Society* 13 (1): 114–133. https://doi.org/10.1177/1461444810365313.
Mason, J. 2012. *Qualitative Researching.* 3rd ed. Sage.
Massimi, M. 2010. Technologies for Supporting Bereaved Families. *Proceedings of the CHI EA'10*: 2951–2954.
Mattingly, Cheryl. 2005. Toward a Vulnerable Ethics of Research Practice. *Health* 9 (4): 453–471.
Moncur, Wendy. 2013. The Emotional Wellbeing of Researchers: Considerations for Practice. *Proceedings of the SIGCHI Conference on Human Factors in Computing Systems (CHI '13).*
Moncur, Wendy, and Miriam Julius. 2014. *Story Shell: The Participatory Design of a Bespoke Digital Memorial*, 1–9.
Phillips, Louise, Marianne Kristiansen, Marja Vehviläinen, and Ewa Gunnarsson. 2013. *Knowledge and Power in Collaborative Research: A Reflexive Approach.* Routledge.
Pitsillides, Stacey. 2017. *Digital Death: The Materiality of Co-Crafted Legacies.* Doctoral dissertation, Goldsmiths, University of London, UK.
Scarth, Bonnie J. 2016. Bereaved Participants' Reasons for Wanting Their Real Names Used in Thanatology Research. *Research Ethics* 12 (2): 80–96. https://doi.org/10.1177/1747016115599569.
Shklovski, Irina, and Janet Vertesi. 2013. Un-Googling Publications: The Ethics and Problems of Anonymization. In *CHI'13 Extended Abstracts on Human Factors in Computing Systems*, 2169–2178. ACM.
Sinatra, Nancy. 1985. *Frank Sinatra My Father.* 1st Edit ed. Garden City, New York: Doubleday & Company.
Smolensky, Kirsten Rabe. 2008. Rights of the Dead. *Hofstra Law Review* 37: 763.
Tilley, Liz, and Kate Woodthorpe. 2011. Is It the End for Anonymity as We Know It? A Critical Examination of the Ethical Principle of Anonymity in the Context of 21st Century Demands on the Qualitative Researcher. *Qualitative Research* 11 (2): 197–212.
Urquhart, Cathy. 2013. *Grounded Theory for Qualitative Research: A Practical Guide.* Sage.
Walford, Geoffrey. 2005. Research Ethical Guidelines and Anonymity. *International Journal of Research & Method in Education* 28 (1): 83–93.

Walter, Tony. 2015. The Dead Who Become Angels: Bereavement and Vernacular Religion. *OMEGA-Journal of Death and Dying*. https://doi.org/10.1177/0030222815575697.

Zimmer, Michael. 2010. But the Data Is Already Public': On the Ethics of Research in Facebook. *Ethics and Information Technology* 12 (4): 313–325.

# Index

**A**
Accidental digital memories and messages, 128, 142
Accidental posthumous site, 43
Anonymity, 170, 171, 173

**B**
Bell, Gordon, 31, 36, 37, 52, 160
Benevolence, 65–72, 80, 154
Bentham, Jeremy, 104, 105, 109, 115, 127, 172
Beyond goodbye, 64
BINA48, 38
Black Mirror, viii, 28, 36, 40, 150, 160
Brooker, Charlie, viii, 28, 36
Brubaker, Jed, 7–11, 42, 48–50, 93, 100, 129, 135, 148–150, 161

**C**
CARL, 117
Charmaz, Kathy, 4, 15, 16, 92, 175
Constructivist grounded theory, 1, 3, 16, 28, 32, 127, 168, 175
Continuing bonds theory, 10, 134–136, 140–142
Control, 2, 7, 13, 19–21, 31, 33–35, 42, 44, 57, 80–83, 93–95, 99, 100, 106–109, 116, 118, 119, 123, 129, 130, 138, 139, 141, 151–153, 155, 156, 159, 160, 162, 178, 179, 182
Covid-19, 3, 14, 55, 117–119

**D**
*Dasein*
  digi-*Dasein*, 56
  digital *Dasein*, 19, 47, 55–58, 129, 140
Death Over Dinner, 14
Digital assets, 4, 34, 35, 154, 177
Digital bonds, 138, 140–142


D. J. Bassett, *The Creation and Inheritance of Digital Afterlives*, Palgrave Studies in the Future of Humanity and its Successors, https://doi.org/10.1007/978-3-030-91684-8

Digital creators
accidental digital creators, 78, 140, 156
intentional digital creators, 6, 27, 41, 47, 50, 57, 61, 63, 65, 78, 128, 131, 140, 142, 150, 156, 157, 159
Digital do not reanimate (DDNR), 20, 128, 156–157, 161, 162
Digital endurance, vii, 4, 5, 156
Digital immortality, 5, 11–12, 30–32, 36, 37, 65, 79–84
Digital inheritors
accidental digital inheritors, 93, 140
intentional digital inheritors, 57, 140
Digital legacy, 4, 33, 141, 151, 154, 177
Digital memories and messages
accidental digital memories and messages, 6, 41, 128, 142, 159, 160
intentional digital memories and messages, 6, 128
Digital *Nachlass*, 56–58
Digital scholarship, 64–65
Digital sociology, 64–65
Digital *v* physical, 109–111
Digital zombies, viii, 1, 19, 21, 30, 31, 33, 37, 38, 40, 48–51, 54, 156, 157, 160, 161
Disenfranchised grief, 9, 118
Do-Not-Resuscitate (DNR), 157
Dual Process Model, 132–134, 137–140

E
ETER9, 36, 38–40, 161
Eternime, 36, 37, 148

F
Facebook, vii, 2, 3, 7–10, 13, 14, 27, 29, 30, 35, 37, 39, 41, 42, 44, 54, 65, 74–76, 78, 79, 93–95, 98–100, 102, 104, 110, 111, 114–117, 127, 129, 130, 147, 149, 150, 183
The Fear of Second Loss, 20, 125–132, 138, 141, 142, 147
Forgetting, 127, 153–156
Right to be forgotten, 147
Freud, Sigmund, 50, 132
Fuller, Steve, 7, 54

G
Gazzard, Hollie, 44, 149
Griefbots, 40
Grief Demystified, 134, 136
Grounded theory, 1, 3, 15–16, 28, 32, 64, 92, 127, 168, 175

H
Heidegger, Martin, 54–56

I
Immortality, 5, 11–12, 19, 30–33, 36, 37, 49, 50, 65, 79–84, 141, 147, 150, 154, 173
Intentional digital memories and messages, 128
Intentional posthumous site, 151
Internet ghosts, 49, 50, 54

K
Kasket, Elaine, 5, 10, 12, 43, 44, 48, 51, 54, 55, 75, 127, 135, 138, 141, 149, 158, 161

Kubler-Ross, Elisabeth, 132, 168
Kuyda, Eugenia, 40

L
Lifelogging, 63, 64, 78, 83
LifeNaut, 36–39, 70
Live streaming funerals, 64, 117, 118, 181–182
Lurking, 10, 20, 62–64

M
Massimi, Michael, 21, 50, 88, 100, 107, 109, 129, 135, 142, 148, 149, 152, 161, 169
Monkhouse, Bob, 49

N
*Nachlass*, 56–58
  digital *Nachlass*, 56–58

P
Pitsillides, Stacey, 4, 148, 161, 171, 173

R
Remembering and forgetting, 127, 153–156
RIP trolling, 19, 43, 179
Rothblatt, Martine, 37, 38, 160

S
SafeBeyond, 34, 35
Schrodinger's cat, 20, 157, 161
Schut, Henk, 90, 132–134, 139, 140
Second death, 123, 127–129
Second loss, 20, 125–132, 138, 141, 142, 147
Service providers (SP)
  accidental service providers, 6, 27–44, 148
  intentional service providers, 6, 19, 27–44, 148, 156, 159, 161
Skype, 8, 20, 28, 48, 87–89, 117, 170
Social media, 1–3, 7, 8, 11, 13–15, 19, 22, 27, 33, 36, 50, 51, 53, 62–67, 69, 73, 76, 77, 79, 107, 111, 116, 177–183
Sociology and death, 20, 123–125
Stage theory, 132, 134, 135, 149
Stokes, Patrick, 47, 52, 53, 111, 112, 125, 127, 128, 157, 160
Stroebe, Margaret, 90, 133, 134, 139, 140

T
Taboo, 12–13, 124
Technophilia, 29, 65, 77–79
Thanablogging, 13, 63, 66–69, 73, 80
Thanabots, viii, 27, 40, 47, 64, 128, 156, 157, 161, 167
Thanatechnology, 5, 8, 9, 11, 19, 21, 48, 50, 56, 57, 61, 74, 75, 88, 90, 106, 111, 128, 132, 140–142, 153, 168, 171, 174, 176
Thanatosensitive design (TSD), 21, 109, 142, 148–150, 152, 161
Therapy, 65, 72–74, 91, 141
Timed posthumous messages, 20, 29, 107–109, 138, 150, 162, 168
Timings, frequency and triggers, 99–107
Tuesdays with Morrie, 41, 170
Twitter, 2, 8, 13, 17, 27, 30, 31, 40, 62, 65, 67, 79, 88, 107, 108, 111, 130, 169, 170, 172

**U**

Up, 4, 5, 8, 13, 18, 29, 32, 33, 36, 37, 39, 42, 43, 68, 74, 78, 79, 82, 93, 96, 97, 103, 105, 107–109, 114, 130, 147, 148, 157, 158, 160, 161, 174, 176, 179

**V**

Victorian mourning jewellery, 130
Voices of the dead, 29, 35, 95–99, 151, 155
Voluntary code of conduct, 21, 44, 142, 158–162

**W**

Walter, Tony, 7–12, 14, 44, 48, 49, 51, 52, 75, 117, 124, 125, 135, 136, 141, 148, 154, 171, 178
Web Sheriff, 149
WhatsApp, 2, 8, 27, 41, 42, 48, 100, 112, 113
World Health Organization, 117

**Y**

YouTube, 8, 13, 33, 41, 43, 44, 62, 72, 115, 125

The manufacturer's authorised representative in the EU is Springer Nature Customer Service Centre GmbH, Europaplatz 3, 69115 Heidelberg, Germany. If you have any concerns regarding our products, please contact ProductSafety@springernature.com

Printed and bound by CPI Group (UK) Ltd, Croydon, CR0 4YY

30/09/2026

02225661-0002